University of Lowell
Alumni/Lydon Library

D0880990

Computational Microelectronics

Edited by S. Selberherr

The Stationary Semiconductor Device Equations

Peter A. Markowich

Springer-Verlag Wien New York

Doz. Dr. Peter A. Markowich
Institut für Angewandte und Numerische Mathematik
Technische Universität Wien, Austria

Printed in Austria

With 40 Figures

ISSN 0179-0307
ISBN 3-211-81892-8 Springer-Verlag Wien-New York
ISBN 0-387-81892-8 Springer-Verlag New York-Wien

Preface

In the last two decades semiconductor device simulation has become a research area, which thrives on a cooperation of physicists, electrical engineers and mathematicians. In this book the static semiconductor device problem is presented and analysed from an applied mathematician's point of view. I shall derive the device equations – as obtained for the first time by Van Roosbroeck in 1950 – from physical principles, present a mathematical analysis, discuss their numerical solution by discretisation techniques and report on selected device simulation runs.
To me personally the most fascinating aspect of mathematical device analysis is that an interplay of abstract mathematics, perturbation theory, numerical analysis and device physics is prompting the design and development of new technology. I very much hope to convey to the reader the importance of applied mathematics for technological progress.
Each chapter of this book is designed to be as selfcontained as possible, however, the mathematical analysis of the device problem requires tools which cannot be presented completely here. Those readers who are not interested in the mathematical methodology and rigor can extract the desired information by simply ignoring details and proofs of theorems. Also, at the beginning of each chapter I refer to textbooks which introduce the interested reader to the required mathematical concepts.
With this book I want to provide device modelers with an overview of mathematical results and discretisation techniques in order to make it possible for them to design more efficient simulation programs and to assess and interpret simulation results profoundly. Also I would like to introduce mathematicians to the semiconductor device problem and encourage them to contribute to this fairly new research area.
I want to express my gratitude to those, who contributed to this book. First of all I thank Richard Weiss, who thaught me applied mathematics and numerical analysis and encouraged me to write this book. I am also extremely grateful to Christian Ringhofer for introducing me to the semiconductor device problem and for many stimulating discussions. I thank Siegfried Selberherr for many hours of discussing device physics and simulation with me. I am also indebted to Wolfgang Agler, Uri Ascher, Frank de Hoog, Andrea Franz, Gerhard Franz, Wolfgang Griebel, Andreas Griewank, Werner Jüngling, Erasmus Langer, Peter Pichler, Hans Pötzl, Christian Schmeiser, Herbert Steinrück, Fritz Straker, Peter Szmolyan, Vidar Thomee, Christoph Überhuber and Milos Zlamal for various

suggestions on the material. I am grateful to John Nohel of the Mathematics Research Center of the University of Wisconsin-Madison for letting me enjoy the stimulating atmosphere at the MRC as often as possible and to Neil Trudinger of the Centre for Mathematical Analysis of the Australian National University for inviting me to the CMA. A large part of the research for Chapter 5 was done there. I thank Ms. Ursula Schweigler for typing skillfully the largest part of the manuscript and Ms. Monika Mazic for proofreading the galleys.
Finally, I want to express my gratitude to my wife Sylvia for her patience while I was working on this book.

Vienna, November 1985

Peter A. Markowich

Contents

Introduction 1

A Review of Semiconductor Device Modeling

Semiconductor device modeling started in the early fifties just after Van Roosbroeck had formulated the so-called fundamental semiconductor device equations, a nonlinear system of partial differential equations, which describes potential distribution, carrier concentrations and current flow in arbitrary semiconductor devices (see [1.32]). In the early stages highly simplified one-dimensional models accessible to direct analytic treatment were used in order to understand device characteristics and to improve device design (see [1.28], [1.29]). The trend towards miniaturisation in VLSI and device design, mainly caused by the increasing demand for fast computers with large storage, rendered the simplified models and consequently the fully analytic approach obsolete. Instead, the emphasis shifted towards numerical simulation techniques, i.e. the computational solution of the semiconductor device equations based on numerical discretisation methods. This approach was suggested by Gummel [1.9] for the bipolar transistor. De Mari [1.3], [1.4] applied the fully computational approach to *pn*-junction diodes. It became clear very soon that standard methods and theories of discretisation techniques are inappropriate because they require an enormous amount of computer resources in order to give reasonably accurate results when modeling practically relevant devices. The main reason for this is that the equations are stiff and allow for solutions of locally different behaviour. The stiffness problem was – to a certain extent – overcome by the ingenuity of Scharfetter and Gummel, who developed a nonstandard, special purpose discretisation method, which – sometimes in a modified and extended way – is being used up to now (see [1.25]).
In the seventies mathematicians started to pay attention to the fundamental semiconductor device equations. At first the emphasis was on proving existence, uniqueness and dependence-on-data results by employing abstract methods from the theory of partial differential equations (see [1.19], [1.20]) and on the 'classical' convergence analysis of discretisation methods (see [1.21], [1.22]). At the same time the understanding of device performance improved and the interaction of mathematical and physical insight enhanced the development of feasable numerical techniques. This made it possible to simulate complicated, intrinsically multi-dimensional devices (see [1.27] for

a detailed list of references). Also the mathematical understanding of the device equations improved and more realistic and complicated device models were rigorously analysed (see, e.g. [1.26]).
Inspired by results from numerical simulations and from device physics an important development took place in the last few years, namely the mathematical exploitation of the stiffness of the device equations by means of singular perturbation theory (see [1.13], [1.14], [1.15], [1.30], [1.31]). The mathematical concept of singular perturbations and matched asymptomic expansions – originating from the investigation of boundary layer problems in fluid dynamics – has recently developed into an important and exciting area of applied mathematics (see, e.g., [1.5], [1.12]). It can be used to analytically determine the structure of solutions of the device problem by distinguishing between regions of fast and slow variation within the device and by describing the local variation of solutions in terms of physical parameters. This analysis provides a mathematically sound basis for the characterisation of depletion layer phenomena, which – from a physical point of view – are well understood.
The singular perturbation analysis of the device problem contributed considerably to the understanding of the performance of discretisation schemes, it prompted the design of highly improved numerical techniques and showed new ways of adaptive meshrefinement (see, e.g., [1.16], [1.24]). This made it possible to design more efficient computer programs, which allow for the simulation of highly miniaturised multilayer devices (see [1.8]). Nowadays virtually every company, which produces semiconductor devices, employs a mixed simulation-prototype construction approach for device design and analysis.
So far a huge amount of papers on device modeling can be found in the engineering and – to a lesser extent – in the mathematical literature. Also, to my knowledge, three monographs on the subject appeared. Kurata presents in his book [1.11] the methods used in the early stages of device modeling, Mock [1.23] focuses on the abstract mathematical results and on a 'classical' convergence analysis of discretisation methods and Selberherr [1.27] extensively treats the physical background and computational techniques. Moreover, international conferences, entirely devoted to computational semiconductor device simulation, were held. The proceedings [1.1], [1.2], [1.6], [1.7], [1.17], [1.18], besides giving a huge amount of information on the subject, serve as an excellent collection of references.

About This Book

Semiconductor device modeling has become an interdisciplinary research area. With this in mind, I tried to make this monograph accessible to mathematicians with interest in applied problems, numerical analysts, device modelers and physicists. I would like to introduce mathematicians to the static semiconductor device problem, make then aware of open problems and provide device modelers with a broad overview of the existing numerical

techniques together with sufficient background information to understand the pitfalls of numerical simulation. The mathematical analysis and the presented numerical results shall point out to physicists the shortcomings of the model and thus motivate them to reassess and improve it.
Electrical phenomena in semiconductor devices have to be described – on a physically rigorous level – by transient models. However, steady state simulations, which require significantly less computer resources, give satisfactory results in many applications. Motivated by this practical aspect and acknowledging the mathematical complexity of the device problem I decided to devote this monograph entirely to the static case.
Each chapter is designed to be as selfcontained as possible. This should enable those readers, who are only interested in certain aspects of the problem, to select the appropriate chapters and skip the others without loosing the context.
Clearly, not all mathematical tools to be used for the analysis can be introduced and explained in detail. Particularly, the analysis presented in Chapter 3 requires abstract methods from the theory of elliptic differential equations. To those readers, who do not want to get involved in this area, I recommend to skip the proofs and just read the theorems and their interpretations. The singular perturbation analysis of Chapter 4 and the discussion of discretisation techniques in Chapter 5 require much less sophisticated mathematical prerequisites. These chapters should be readily accessible to readers with an engineering-type mathematics education.
Chapter 2, particularly devoted to mathematicians, is concerned with the derivation of the mathematical formulation of the device problem from basic physical principles and with the modeling of physical parameters. Simulation results are presented in Chapter 6.
A list of mathematical notations and symbols is compiled in the Appendix.

Semiconductors

For those readers, who are not at all familiar with solid state physics, I shall here explain – in a highly simplistic way – the basic properties of semiconductors. Detailed information can be obtained from [1.28], [1.29].
Often solids are classified as either insulators, semiconductors or conductors. At room temperature the electrical conductivity of a semiconductor is significantly larger than the conductivity of an insulator and it is significantly smaller than the conductivity of a conductor. This can be explained by comparing the intrinsic concentrations of conduction electrons of the three types of solids. For (the technologically most relevant) semiconductor silicon at room temperature it approximately equals 10^{10} cm^{-3}, for a typical conductor, e.g. a metal, it is of the order of magnitude 10^{22} cm^{-3} and in the typical insulator diamond there are only a few thousand conduction electrons per cm^3. When the energy level of a semiconductor crystal is moderately raised by, e.g., applying an electric field or increasing the temperature, then relatively many valence electrons, originally responsible for the chemical

compound, become conduction electrons. Clearly, every electron, which has changed its state in this way, leaves a 'hole' in the lattice. It turns out that it is reasonable to regard these holes as positively charged carriers in semiconductors. The movement of holes causes – just as the movement of the negatively charged conduction electrons – an electrical current.
For a semiconductor in thermal equilibrium the number of holes equals the number of conduction electrons. In the sequel we shall denote this intrinsic carrier concentration by n_i. Detailed information on charged carriers and on current flow can be found in [1.10]. The properties of the intrinsic carrier concentration are discussed in [1.27].
The most important step in the fabrication of a semiconductor device is the controlled implantation of impurity atoms into a semiconductor crystal. This process is usually called 'doping'. Principally, two possibilities exist. Either dopant atoms, which can produce one or more excess conduction electrons (called donors), or dopant atoms, which can accept electrons and thus 'produce' holes (called acceptors), can be implanted into the crystal. It is technologically possible to implant dopants of a concentration, which is several orders of magnitude larger than the intrinsic carrier concentration n_i, into semiconductor crystals. This increases the conductivity significantly. Thus, the electrical properties of the crystal can be controlled by doping.
The performance of a semiconductor device is mainly determined by the distributions of donors and acceptors. As an example we depict (a simplified

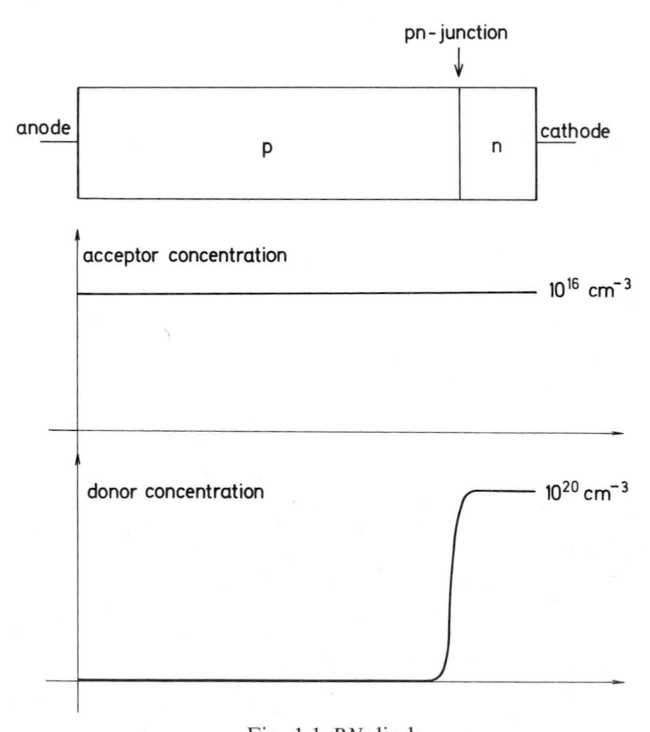

Fig. 1.1 *PN*-diode

model of) the most basic semiconductor device, namely the *pn*-junction diode, in Fig. 1.1. The left '*p*-side', doped with acceptors, is positively charged and the right '*n*-side', predominantly doped with donors, is negatively charged. The anode and cathode contacts are fabricated by bringing a metal into intimate contact with the doped semiconductor crystal.
When a positive bias is applied to the anode-cathode contacts, a large current flows through the diode, even if the applied voltage is small. For negative applied voltages only small (leakage) currents occur. Thus the *pn*-diode acts as a 'valve', it 'opens' and 'closes' depending on the sign of the applied voltage.
An excellent performance analysis of the *pn*-diode and of more complicated, technologically relevant multiple-junction and multiple-contact devices can be found in [1.29].

References

[1.1] Browne, B. T., Miller, J. J. H. (eds.): Numerical Analysis of Semiconductor Devices. Dublin: Boole Press 1979.
[1.2] Browne, B. T., Miller, J. J. H. (eds.): Numerical Analysis of Semiconductor Devices and Integrated Circuits. Dublin: Boole Press 1981.
[1.3] De Mari, A.: An Accurate Numerical Steady State One-Dimensional Solution of the P–N Junction. Solid State Electron. *11,* 33–58 (1968).
[1.4] De Mari, A.: An Accurate Numerical One-Dimensional Solution of the P–N Junction under Arbitrary Transient Conditions. Solid State Electron. *11,* 1021–2053 (1968).
[1.5] Eckhaus, W.: Asymptotic Analysis of Singular Perturbations. Amsterdam–New York–Oxford: North-Holland 1979.
[1.6] Fichtner, W., Rose, D. (eds.): Special Issue on Numerical Simulation of VLSI Devices. IEEE Trans. Electron Devices. *ED–30* (1983).
[1.7] Fichtner, W., Rose, D. (eds.): Special Issue on Numerical Simulation of VLSI Devices. IEEE Trans. Electron Devices. (To appear.)
[1.8] Franz, A. F., Franz, G. A., Selberherr, S., Ringhofer, C., Markowich, P.: Finite Boxes – A Generalisation of the Finite Difference Method Suitable for Semiconductor Device Simulation. IEEE Trans. Electron Devices. *ED–30,* No. 9, 1070–1082 (1983).
[1.9] Gummel, H. K.: A Self-Consistent Iterative Scheme for One-Dimensional Steady State Transistor Calculations. IEEE Trans. Electron Devices. *ED–11,* 455–465 (1964).
[1.10] Heywang, W., Pötzl, H. W.: Bandstruktur und Stromtransport. Berlin–Heidelberg–New York: Springer 1976.
[1.11] Kurata, M.: Numerical Analysis for Semiconductor Devices. Lexington, Mass.: Lexington Press 1982.
[1.12] O'Malley, R. E., jr.: Introduction to Singular Perturbations. New York: Academic Press 1974.
[1.13] Markowich, P. A., Ringhofer, C.: A Singularly Perturbed Boundary Value Problem Modelling a Semiconductor Device. SIAM J. Appl. Math. *44,* No. 2, 231–256 (1984).
[1.14] Markowich, P. A.: A Singular Perturbation Analysis of the Fundamental Semiconductor Device Equations. SIAM J. Appl. Math. *44,* No. 5, 896–928 (1984).
[1.15] Markowich, P. A.: A Qualitative Analysis of the Fundamental Semiconductor Device Equations. COMPEL *2,* No. 3, 97–115 (1983).
[1.16] Markowich, P. A., Ringhofer, C., Selberherr, S., Lentini, M.: A Singular Perturbation Approach for the Analysis of the Fundamental Semiconductor Equations. IEEE Trans. Electron Devices. *ED–30,* No. 9, 1165–1180 (1983).
[1.17] Miller, J. J. H. (ed.): Numerical Analysis of Semiconductor Devices and Integrated Circuits. Dublin: Boole Press 1983.

[1.18] Miller, J. J. H. (ed.): Numerical Analysis of Semiconductor Devices and Integrated Circuits. Dublin: Boole Press 1985.
[1.19] Mock, M. S.: On Equations Describing Steady State Carrier Distributions in a Semiconductor Device. Comm. Pure and Appl. Math. *25,* 781–792 (1972).
[1.20] Mock, M. S.: An Initial Value Problem from Semiconductor Device Theory. SIAM J. Math. Anal. *5,* No. 4, 597–612 (1974).
[1.21] Mock, M. S.: Time Discretisation of a Nonlinear Initial Value Problem. J. Comp. Phys. *21,* 20–37 (1976).
[1.22] Mock, M. S.: On the Computation of Semiconductor Device Current Characteristics by Finite Difference Methods. J. Eng. Math. *7,* No. 3, 193–205 (1973).
[1.23] Mock, M. S.: Analysis of Mathematical Models of Semiconductor Devices. Dublin: Boole Press 1983.
[1.24] Ringhofer, C., Selberherr, S.: Implications of Analytical Investigations about the Semiconductor Equations on Device Modelling Programs. MRC–TSR 2513, Math. Res. Center, University of Wisconsin-Madison, U.S.A., 1980.
[1.25] Scharfetter, D. L., Gummel, H. K.: Large-Signal Analysis of a Silicon Read Diode Oscillator. IEEE Trans. Electron Devices. *ED–16,* 64–77 (1969).
[1.26] Seidman, T. I.: Steady State Solutions of Diffusion-Reaction Systems with Electrostatic Convection. Nonlinear Analysis. *4,* No. 3, 623–637 (1980).
[1.27] Selberherr, S.: Analysis and Simulation of Semiconductor Devices. Wien–New York: Springer 1984.
[1.28] Smith, R. A.: Semiconductors, 2nd ed. Cambridge: Cambridge Univ. Press 1978.
[1.29] Sze, S. M.: Physics of Semiconductor Devices, 2nd ed. New York: J. Wiley 1981.
[1.30] Vasileva, A. B., Stelmakh, V. F.: Singularly Disturbed Systems of the Theory of Semiconductor Devices. USSR Comput. Math. Phys. *17,* 48–58 (1977).
[1.31] Vasileva, A. B., Butuzow, V. F.: Singularly Perturbed Equations in the Critical Case. MRC–TSR 2039, Math. Res. Center, University of Wisconsin-Madison, U.S.A., 1980.
[1.32] van Roosbroeck, W. V.: Theory of Flow of Electrons and Holes in Germanium and Other Semiconductors. Bell Syst. Techn. J. *29,* 560–607 (1950).

Mathematical Modeling of Semiconductor Devices 2

In this Chapter we shall formulate the system of partial differential equations, which describes potential distribution, carrier concentrations and current flow in semiconductor devices. We shall supplement the system by boundary conditions representing the interaction of the device with the outer world and discuss the modeling of physical parameters appearing in the system. Also, various choices of dependent variables, which are useful for analytical purposes, will be presented. Finally, we shall scale the physical quantities and put the system of equations and the boundary conditions into a dimensionless form appropriate for further mathematical and numerical investigations.

2.1 Derivation of the Basic Semiconductor Device Equations

We shall now derive the basic mathematical model for the electrodynamic behaviour of semiconductor devices.

A semiconductor device occupies a bounded, simply connected domain in $\mathbb{R}^3$, which we denote by Λ. It consists of a semiconductor part occupying the subdomain Ω of Λ, and – in the case of a *m*etal-*o*xide-*s*emiconductor (MOS) device – of one or more thin adjacent oxide-domains, whose union we denote by Φ.

For a pure semiconductor device $\Lambda = \Omega$ and $\Phi = \{\}$ holds.

The model equations are partly based on:

Maxwell's Equations

The evolution of the electromagnetic field quantities in an arbitrary medium is governed by Maxwell's equations (see, e.g., [2.8], [2.20]):

$$\operatorname{rot} H = \mathrm{J} + \frac{\partial D}{\partial t}, \tag{2.1.1}$$

$$\operatorname{rot} E = -\frac{\partial B}{\partial t}, \tag{2.1.2}$$

$$\operatorname{div} D = \varrho, \tag{2.1.3}$$

$$\operatorname{div} B = 0. \tag{2.1.4}$$

E and D are the electric field and displacement vectors resp., H and B the magnetic field and induction vectors resp., J is the conduction current density and ϱ the electric charge density. By $x = (x_1, x_2, x_3) \in \mathbb{R}^3$ we shall denote the (independent) space variable and by $t \geqq 0$ the time variable.
For the following we relate the electric field and the electric displacement by

$$D = \varepsilon E, \tag{2.1.4) (a}$$

where ε is the permittivity of the medium. We assume that ε is time independent and spatially homogeneous in the semiconductor as well as in possibly occurring oxide regions of the device. Principally, ε is a 3×3-matrix. We, however, assume that the material is isotropic and, thus, treat ε as a scalar. These assumptions are realistic for most common applications of semiconductor devices (see [2.16]). Numerical values for ε for various materials are given in Table 2.1.1.

Table 2.1.1. Numerical values of physical constants

Symbol	Meaning	Value
q	elementary charge	$1.6021892 \times 10^{-19}$ As
ε_v	permittivity constant in vacuum	$8.854187818 \times 10^{-14}$ AsV^{-1} cm^{-1}
c	speed of light in vacuum	$2.99792458 \times 10^{10}$ cms^{-1}
k_B	Boltzmann's constant	$1.3806622 \times 10^{-23}$ VAsK^{-1}

Permittivity constants

Material	*Value*
silicon	$11.7\ \varepsilon_v$
silicon-dioxide	$3.9\ \varepsilon_v$
germanium	$16.1\ \varepsilon_v$

Poisson's Equation

The equation (2.1.4) can be satisfied by the introduction of a vector potential:

$$B = \operatorname{rot} A. \tag{2.1.5}$$

Clearly, for given magnetic induction vector B, the vector potential A is not uniquely defined. Therefore, we shall specify the divergence of A later on. Inserting (2.1.5) into (2.1.2) gives

$$\operatorname{rot}\left(E + \frac{\partial A}{\partial t}\right) = 0. \tag{2.1.6}$$

Since a sufficiently smooth vortex-free vector field, which is defined in a simply connected domain, is a gradient field, we have:

$$E = -\frac{\partial A}{\partial t} - \text{grad}\,\psi \tag{2.1.7}$$

for some scalar potential ψ. From (2.1.4) (a) and (2.1.3) we obtain

$$\varepsilon \frac{\partial}{\partial t} \text{div}\, A + \varepsilon \Delta \psi = -\varrho. \tag{2.1.8}$$

In order to make the equation invariant under the Lorentz transformation (see [2.8]) we set:

$$\text{div}\, A = -\frac{1}{c^2}\frac{\partial \psi}{\partial t} \quad \text{(Lorentz convention)}, \tag{2.1.9}$$

where c denotes the speed of light (see Table 2.1.1).
We obtain the wave equation:

$$-\frac{\varepsilon}{c^2}\frac{\partial^2 \psi}{\partial t^2} + \varepsilon \Delta \psi = -\varrho. \tag{2.1.10}$$

Usually, it is assumed that the speed of light is large compared to characteristic propagation velocities in the device. Therefore the wave-propagation term is neglected and Poisson's equation for the electrical potential ψ is obtained:

$$\varepsilon \Delta \psi = -\varrho. \tag{2.1.11}$$

In the semiconductor we can write the space charge density ϱ as:

$$\varrho = q(p - n + C), \qquad x \in \Omega, \tag{2.1.12}$$

where q is the elementary charge as given in Table 2.1.1, p the concentration of (positively charged) holes; n the concentration of (negatively charged) conduction electrons and C the predefined 'profile' of electrically active deposits. Under the usual assumption that all impurity atoms are singly ionised we have:

$$C = N_D^+ - N_A^-. \tag{2.1.13}$$

N_D^+ denotes the concentration of electrically active donor and N_A^- the concentration of electrically active acceptor atoms.
Usually, the oxide is assumed to be charge-neutral:

$$\varrho = 0, \qquad x \in \Phi. \tag{2.1.14}$$

Comprising the results, we write Poisson's equation in the form commonly used for device simulation:

$$\varepsilon \Delta \psi = \begin{cases} q(n - p - C), & x \in \Omega \\ 0, & x \in \Phi \end{cases}. \tag{2.1.15}$$

Continuity Equations

Applying the divergence operator to (2.1.1) and using (2.1.3) gives

$$0 = \operatorname{div} J + \frac{\partial \varrho}{\partial t}, \tag{2.1.16}$$

since divrot $H = 0$ for every sufficiently differentiable vectorfield H. Sources and sinks of the conduction current density are solely determined by the temporal variation of the charge density.

We split the conduction current density J into a part J_n caused by electrons and a part J_p caused by holes:

$$J = J_n + J_p. \tag{2.1.17}$$

We call J_n the electron current density and J_p the hole current density. For the following we assume that the doping profile is time-invariant:

$$\frac{\partial C}{\partial t} = 0, \tag{2.1.18}$$

i.e. we neglect charge defects (see [2.16]).

By using (2.1.17), (2.1.18) and (2.1.12) we derive:

$$-\operatorname{div} J_p - q\frac{\partial p}{\partial t} = \operatorname{div} J_n - q\frac{\partial n}{\partial t}, \qquad x \in \Omega. \tag{2.1.19}$$

We obtain an equation for the electron current density and an equation for the hole current density by setting both sides of (2.1.19) equal to a quantity, which we write as qR:

$$\operatorname{div} J_n - q\frac{\partial n}{\partial t} = qR, \qquad x \in \Omega, \tag{2.1.20}$$

$$\operatorname{div} J_p + q\frac{\partial p}{\partial t} = -qR, \qquad x \in \Omega. \tag{2.1.21}$$

Clearly, no new information is obtained by rewriting (2.1.19) as two equations. However, by inspection of the left hand side of (2.1.20), (2.1.21), the quantity R can physically be interpreted as the difference of the rate at which electron-hole carrier pairs recombine and the rate at which they are generated in the semiconductor. Therefore we call R the recombination-generation rate. Clearly, generation prevails in those subdomain of Ω, in which

$$R < 0 \tag{2.1.22 (a)}$$

holds and recombination prevails if

$$R > 0. \tag{2.1.22 (b)}$$

Further information on R has to be obtained by the methods of statistical physics solely using the interpretation as recombination-generation rate, and, as will be discussed later on, (more or less) appropriate models exist.

When a mathematical model for R is inserted into (2.1.20), (2.1.21), each of these two equations has a meaning of its own, and the purely formal step of splitting the continuity equation (2.1.16) for the total conduction current density into two separate continuity equations for the electron and for the hole current density resp. gets mathematical and physical significance.
The oxide is assumed to be an ideal insulator, therefore $J_n|_\Phi \equiv J_p|_\Phi \equiv 0$.

Current Relations

The derivation of equations, which express the current densities J_n and J_p in terms of the electric field and the carrier concentrations, is an extremely cumbersome task. Usually, Boltzmann's transport equation (see [2.24]) constitutes the starting point and rather simple current relations are obtained by lengthy calculations, which are facilitated by a lot of physically highly restrictive assumptions. Interesting overviews of this approach are given in the references [2.16], [2.17].
We shall, however, not follow these lines, but instead present a purely phenomenological derivation of the current relations. Therefore, we identify the two main sources for current flow in semiconductor devices:

(a) diffusion of the electron and hole ensembles with resulting diffusion current densities J_n^{diff}, J_p^{diff},

(b) drift of electrons and holes caused by the electric field as driving force with resulting drift current densities J_n^{drift}, J_p^{drift}.

The principal assumption to be used is that the electron and hole current flows are determined by linearly superimposing the diffusion and the drift processes, i.e.:

$$J_n = J_n^{\text{diff}} + J_n^{\text{drift}}, \qquad J_p = J_p^{\text{diff}} + J_p^{\text{drift}}. \tag{2.1.23}$$

Electrons and holes diffuse from regions of high concentration into regions of low concentration. The direction of diffusion of a particle ensemble is the direction of steepest descent of the corresponding particle concentration, and by Fourier's law, the diffusion flux densities are proportional to the gradients of the corresponding particle concentration (see [2.11]). The diffusion current densities are obtained by multiplying the diffusion fluxes with the charge per particle, which is $-q$ for electrons and $+q$ for holes:

$$J_n^{\text{diff}} = qD_n \operatorname{grad} n \qquad (2.1.24)\ (a)$$

$$J_p^{\text{diff}} = -qD_p \operatorname{grad} p \qquad (2.1.24)\ (b)$$

The signs of the right hand sides are chosen such that the diffusion coefficients D_n and D_p are positive.
The electric field driven drift current densities are defined as the products of the charge per particle, the corresponding carrier concentration and the average drift velocity, denoted by v_n^d for electrons and v_p^d for holes:

$$J_n^{\text{drift}} = qnv_n^d \tag{2.1.25 (a)}$$

$$J_p^{\text{drift}} = qpv_p^d \tag{2.1.25 (b)}$$

The drift directions of the carriers are assumed to be parallel to the electric field, the drift of holes has the same orientation as the electric field, while the drift of electrons has opposite orientation. The drift velocities are proportional to the electric field at moderate field strengths (see [2.20]):

$$v_n^d = -\mu_n E, \quad v_p^d = \mu_p E, \tag{2.1.26}$$

where the positive coefficients μ_n, μ_p are called electron and hole mobility resp.
By inserting (2.1.26) into (2.1.25) and by using (2.1.24), (2.1.23) we obtain the current relations:

$$J_n = qD_n \operatorname{grad} n + q\mu_n nE, \quad x \in \Omega. \tag{2.1.27}$$

$$J_p = -qD_p \operatorname{grad} p + q\mu_p pE, \quad x \in \Omega. \tag{2.1.28}$$

Other sources for current flow, important for some applications, stem from temperature variation and variation of the intrinsic carrier concentration within the device (see [2.16], [2.17]). Since these drift current contributions only have a minor impact on the mathematical treatment of the device problem we shall neglect them in the sequel.
Usually, the diffusion coefficients D_n and D_p are related to the mobilities μ_n, μ_p by Einstein's relations:

$$D_n = U_T \mu_n, \quad D_p = U_T \mu_p, \tag{2.1.29}$$

where U_T stands for the thermal voltage given by:

$$U_T = \frac{k_B T}{q}. \tag{2.1.30}$$

k_B denotes Boltzmann's constant (see Table 2.1.1) and T the device temperature.
Einstein's relations are obtained as approximations by methods of statistical physics (see [2.19], [2.20]). The error introduced by them is acceptably small, if the semiconductor is nondegenerate and if the device temperature is 'nearly' spatially homogeneous (see [2.20]).
Clearly, the current relations (2.1.27), (2.1.28) are not derived from physical principles as well established as, for example, Maxwell's equations. The linear superimposition of convective and diffusive current density terms is a mathematically convenient constitutive assumption, which yields reasonably simple expressions for the current densities. However, the current relations should not be accepted in the same way as the continuity equations and Poisson's equation, their validity has to be investigated by other means.
The more rigorous approach to the derivation of current relations outlined in the beginning of this paragraph leads to the same expressions but it gives some insight into the physical assumptions necessary for their validity. We refer the interested reader to the references [2.16], [2.17].

Summary

We shall now summarise the results and state the equations commonly used for device simulation, often called the basic semiconductor device equations, in the form to be used for all further investigations. To achieve a trade-off between the complexity of the physical 'real' world and structural simplicity of the equations, we make the following assumptions:

- The only sources for current flow are electric field driven convection and diffusion of particle ensembles.
- Einstein's relations (2.1.29) hold.
- The magnetic inductions vector B is independent of time.

Therefore the vector potential A is independent of time, and (2.1.7) implies that the electric field equals the negative gradient of the potential ψ:

$$E = -\operatorname{grad} \psi. \tag{2.1.31}$$

Under these assumptions the following system of partial differential equations is obtained:

$$\left.\begin{array}{lll}
\varepsilon \Delta \psi = q(n-p-C) & \text{Poisson's equation} & (2.1.32)\\
\operatorname{div} J_n - q\dfrac{\partial n}{\partial t} = qR & \text{electron continuity equation} & (2.1.33)\\
\operatorname{div} J_p + q\dfrac{\partial p}{\partial t} = -qR & \text{hole continuity equation} & (2.1.34)\\
J_n = q\mu_n(U_T \operatorname{grad} n - n \operatorname{grad} \psi) & \text{electron current relation} & (2.1.35)\\
J_p = -q\mu_p(U_T \operatorname{grad} p + p \operatorname{grad} \psi) & \text{hole current relation} & (2.1.36)
\end{array}\right\} \quad x \in \Omega, \quad t > 0.$$

For MOS-devices Laplace's equation holds in the oxide:

$$\Delta \psi = 0, \qquad x \in \Phi, \qquad t > 0. \tag{2.1.37}$$

The basic semiconductor device equations were for the first time derived by Van Roosbroeck [2.21] in almost this form.
The device temperature T is an internal quantity, which depends strongly on the current densities. Thus, when a rigorous modeling approach is desired, the basic semiconductor device equations (2.1.32)–(2.1.37) have to be supplemented by a heat flow equation, which determines T in dependence of J_n, J_p and certain material data (see [2.16]). For the sake of brevity and transparency we shall, however, for the following assume that T is an externally defined (positive) constant. We remark that T is an input parameter for many device simulation codes, since the numerical solution of the heat flow equation requires an excess use of computer resources, which – in most practical applications – cannot be justified by the slightly more realistic quantitative results.
Clearly, the device temperature must be modeled as an internal quantity when thermal breakdown phenomena shall be investigated.

2.2 Parameter Models

The modeling of the physical parameters of the basic semiconductor device equations, based on modern solid state physics, is an extremely difficult subject and to a large extent not fully explored yet. It has to be done with extreme care when quantitative results as accurate as possible are desired in numerical simulations. For our exploratory purposes, however, it suffices to get a basic understanding of the types of models, which are commonly used, in order to be able to physically interpret and assess the mathematical assumptions employed in the subsequent analysis.
In this section we shall therefore only collect the most important features of the parameter models. We refer those readers, who want to get deeper into the physics of parameter modeling, to the book [2.16], which gives an excellent state of the art report.

The Doping Profile

The performance of a semiconductor device is mainly determined by its doping profile. In the following chapters we shall mathematically exploit certain qualitative and quantitative properties of the doping profile in order to derive structural information on the potential, carrier concentrations and current densities. Therefore we shall now describe the basic structure of physical doping profiles. Details can be found in [2.16], [2.1], [2.2], [2.5].
The most common doping technique is ion implantation, i.e. a highly energized beam of ions penetrates the semiconductor through one of its surfaces. In this way a certain distribution of impurity atoms in the semiconductor crystal is obtained. Of course, more than one implantation process is necessary for the fabrication of multi-layer semiconductor devices.
After ion implantation the impurity atoms have to be redistributed in the semiconductor lattice in order to obtain desired shapes of the donor and acceptor concentrations and in order to well incorporate and ionise 'misplaced' dopant atoms. The physical mechanism responsible for this redistribution is a diffusion process with superimposed electric field driven convection. In practice this process is heat-stipulated. It is performed in a diffusion furnace over a certain time period, which has to be adjusted to the desired shapes of the dopant concentrations.
The doping profile, denoted by $C(x)$ in the sequel, is defined as the net concentration of all ionised impurities, i.e. it is the difference of the electrically active donor concentration and the electrically active acceptor concentration.
We call a subdomain Ω_1 of the semiconductor domain Ω an n-domain, if the concentration of electrically active donors exceeds the concentration of electrically active acceptors in Ω_1, i.e. if $C(x) > 0$ for all $x \in \Omega_1$, and we call Ω_1 a p-domain, if the concentration of electrically active acceptors exceeds the concentration of electrically active donors in Ω_1, i.e. if $C(x) < 0$ for all $x \in \Omega_1$. A joint boundary of an n- and a p-domain is called a pn-junction, a

joint 'boundary' of a highly doped n-domain with a moderately doped n-domain is called an n^+n-junction. p^+p-junctions are defined accordingly. To illustrate the qualitative properties of realistic doping profiles we present a characteristic profile of a MOS-transistor in Fig. 2.2.1. The doping is assumed to be homogeneous in the third spatial dimension over a large region of the device (see Figs. 2.3.1, 2.3.2 for the geometry).

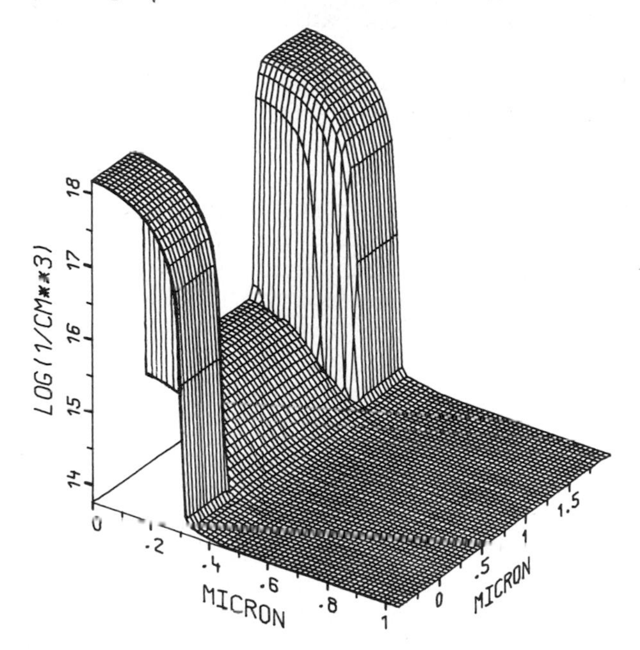

Fig. 2.2.1 Doping profile of a MOS-transistor [log.]

It is clearly visible that the doping profile varies extremly fast close to (the two) pn-junctions while it varies slowly away from the junctions. This property of realistic doping profiles will later on be used to analyse the structure of solutions of the basic semiconductor device equations.
Motivated by the occurence of large gradients of doping profiles close to device junctions we shall for reasons of mathematical convenience and transparency occasionally assume that device junctions are abrupt, i.e. the doping profile is assumed to have jump-discontinuities everywhere on the junctions.
Clearly, abrupt doping profiles are unphysical, but they provide good structural approximations of moderately diffused doping profiles (see [2.20]) and they are extremely advantageous for the structural analysis of solutions of the basic semiconductor device equations.
Moreover, in order to control the electrical behaviour of semiconductor devices by the impurities, it is necessary that the maximal doping concentration

$$C_{\max} := \max_{\Omega} |C|$$

is significantly larger than the intrinsic carrier concentration of the semiconductor at the operating temperature. For the doping profile of Fig. 2.2.1 we have $C_{\max} \approx 2 \times 10^{18}\,\mathrm{cm}^{-3}$, while the intrinsic carrier concentration $n_i \approx 10^{10}\,\mathrm{cm}^{-3}$ for silicon at room temperature.

The Recombination-Generation Rate

For the derivation of the continuity equations we introduced the difference R of the rates of recombination and generation of electron-hole carrier pairs. Recombination occurs when a conduction electron becomes a valence electron and neutralises a hole. Generation occurs when a valence electron becomes a conduction electron and leaves a hole. Generation requires energy and recombination sets energy free.
When a (doped) semiconductor is in thermal equilibrium, then there is a dynamic balance between the recombination and the generation rates. Thus $R \equiv 0$ holds in thermal equilibrium. The equilibrium carrier concentrations n_e and p_e are related by the mass-action law:

$$n_e p_e = n_i^2. \tag{2.2.1}$$

When the thermal equilibrium is disturbed externally by, say, applying a bias to the semiconductor, then the carrier concentrations depart from their equilibrium values and various recombination-generation processes, which try to restore the system to equilibrium, occur. Thus, recombination dominates, i.e. $R > 0$, when excess carriers were generated and generation dominates, i.e. $R < 0$, when carriers were removed.
Various energy transition processes are responsible for the occurrence of recombination-generation. Each of these processes is modeled by a corresponding recombination-generation term, obtained by quantum physical considerations (see [2.16] for details). The most basic recombination-generation process, namely two-particle-transition, is described by the Shockley–Read–Hall term:

$$R_{SRH} = \frac{np - n_i^2}{\tau_p^l(n+n_i) + \tau_n^l(p+n_i)} \tag{2.2.2}$$

(see [2.7], [2.18]). τ_n^l and τ_p^l are the electron and hole life-times resp. They are usually modeled as doping dependent functions. Typical values for τ_p^l and τ_l^n are given in Table 2.2.1.
Three particle transition is modeled by the Auger recombination-generation term:

$$R_{AU} = (C_n^{AU} n + C_p^{AU} p)\,(np - n_i^2) \tag{2.2.3}$$

(see Table 2.2.1 for numerical values of the coefficients C_n^{AU}, C_p^{AU}).
A pure generation process, called impact ionisation, is extremly significant at high electric fields. The impact ionisation generation rate, also called avalanche generation rate, reads:

$$R_I = -\alpha_n \frac{|J_n|}{q} - \alpha_p \frac{|J_p|}{q}, \tag{2.2.4}$$

where the coefficients $\alpha_n > 0$, $\alpha_p > 0$ are the ionisation rates for holes and electrons. They depend strongly on the corresponding electric field components in direction of the current flows:

$$E_n = \frac{E \cdot J_n}{|J_n|}, \qquad E_p = \frac{E \cdot J_p}{|J_p|}. \tag{2.2.5}$$

A variety of suggestions for modeling the ionisation rates exists in the literature (see [2.15]). Well accepted models are:

$$\alpha_n = \alpha_n^\infty \exp\left(-\frac{E_n^{\mathrm{crit}}}{|E_n|}\right), \qquad \alpha_p = \alpha_p^\infty \exp\left(-\frac{E_p^{\mathrm{crit}}}{|E_p|}\right). \tag{2.2.6}$$

Numerical values for the coefficients can be found in Table 2.2.1. In simplified avalanche models the electric field strength $|E|$ is substituted for $|E_n|$ and $|E_p|$.
Other, less important recombination-generation mechanism, like optical and surface recombination-generation processes, are omitted here (see [2.16]).
Usually, the described processes are linearly superimposed and possibly occurring nonlinear interaction is ignored:

$$R = R_{SRH} + R_{AU} + R_I. \tag{2.2.7}$$

For some device simulations, however, one or more of these processes can be neglected for reasons of computational expedience without a major setback of reality. Generally Auger and impact ionisation processes are only significant, when high injection conditions shall be investigated. In particular, impact ionisation is relevant, if at least one junction of the device is reverse biased (the potential drop at the junction is larger than in thermal equilibrium) and Auger recombination-generation has to be taken into account if high carrier concentrations occur, that is if at least one junction is forward biased (the potential drop at the junction is smaller than in thermal equilibrium).

Table 2.2.1. Typical recombination-generation model data for silicon at room temperature

Symbol	*Value*	
τ_n^l	10^{-6} s	
τ_p^l	10^{-5} s	Shockley-Read-Hall Model
n_i	10^{10} cm^{-3}	
C_n^{AU}	2.8×10^{-31} cm^6 s^{-1}	Auger Model
C_p^{AU}	9.9×10^{-32} cm^6 s^{-}	
α_n^∞	10^6 cm^{-1}	
α_p^∞	2×10^6 cm^{-1}	Avalanche Model
E_n^{crit}	1.66×10^6 Vcm^{-1}	
E_p^{crit}	2×10^6 Vcm^{-1}	

We shall in the sequel also prove results under the assumption of a vanishing recombination-generation rate, i.e. $R \equiv 0$. Clearly, this is only physically reasonable for conditions close to thermal equilibrium, however, it is extremely helpful for conceptual considerations since it allows for a transparent mathematical analysis.

Carrier Mobility Models

The carrier mobilities μ_n and μ_p were introduced in the relations (2.1.26) (for the drift velocities) as proportionality factors of the electric field. Physically, they relate to the electron and hole relaxation times τ_n^r and τ_p^r by:

$$\mu_n = \frac{q\tau_n^r}{m_n^*}, \qquad \mu_p = \frac{q\tau_p^r}{m_p^*}, \tag{2.2.8}$$

where m_n^* and m_p^* are the effective electron and hole masses resp.
The relaxation times represent the average time between two consecutive scattering events of carriers. Therefore, to model the mobilities accurately, the scattering mechanisms have to be identified and described properly. Usually, lattice scattering is accounted for by purely temperature dependent mobility models, ionised impurity scattering by temperature and doping dependent models and carrier-carrier scattering by carrier concentration dependent models. Combinations of scattering effects are described by inverse averages of the corresponding single effect models. Details can be found in [2.16].
We remark that the commonly used models are not based on sound physical principles, they were derived by fits to experimental data.
Typically, the electron mobility varies between $50\,\text{cm}^2\,\text{V}^{-1}\,\text{s}^{-1}$ and $1500\,\text{cm}^2\,\text{V}^{-1}\,\text{s}^{-1}$ while the hole mobility varies between $50\,\text{cm}^2\,\text{V}^{-1}\,\text{s}^{-1}$ and $500\,\text{cm}^2\,\text{V}^{-1}\,\text{s}^{-1}$ (for silicon at room temperature).
In semiconductor device physics it is well known that the linear proportionality of the electric field strength and the magnitudes of the drift velocities as formally implied by (2.1.26) only holds at relatively low electric fields. The drift velocities saturate at high electric fields due to carrier heating, i.e.:

$$\lim_{|E|\to\infty} |v_n^d| = v_s^n, \qquad \lim_{|E|\to\infty} |v_p^d| = v_s^p \tag{2.2.9}$$

holds with so called saturation velocities v_s^n and v_s^p. Because of

$$|v_n^d| = \mu_n |E|, \qquad |v_p^d| = \mu_p |E|, \tag{2.2.10}$$

this effect must be accounted for by field-dependent mobilities, if high field effects are to be analysed.
Appropriate models are given in [2.16]. We only mention:

$$\mu_n^{s,E} = \frac{\mu_n^s}{1+\left(\dfrac{\mu_n^s |E|}{v_s^n}\right)}, \qquad \mu_p^{s,E} = \frac{\mu_p^s}{1+\left(\dfrac{\mu_p^s |E|}{v_s^p}\right)}, \tag{2.2.11}$$

where μ_n^s, μ_p^s stand for one of the field-independent scattering mobility models.
The saturation velocities depend weakly on the temperature. Numerical data are given in [2.16].
Inhomogeneous field-independent mobility models do not complicate the largest part of the mathematical analysis of the semiconductor device equations. The inclusion of velocity saturation effects, though physically very important, introduces a great deal of mathematical difficulties, particularly since in this case the functions modeling the mobilities are not bounded away from zero uniformly as $|E| \to \infty$.

2.3 Boundary Conditions and Device Geometry – The Static Problem

We shall now supplement the basic semiconductor device equations by physically appropriate boundary conditions modeling the interaction of devices with their environment. Also we shall discuss the reduction of the transient device problem to the static case as well as the reduction to two or even one space dimension facilitated by geometric properties of certain devices.

Boundary Conditions

The boundary $\partial \Lambda$ of a semiconductor device can generally be split into two disjoint parts:

$$\partial \Lambda = \partial \Lambda_P \cup \partial \Lambda_A, \qquad \partial \Lambda_P \cap \partial \Lambda_A = \{\}.$$

$\partial \Lambda_P$ constitutes a real physical boundary consisting of insulating segments covered by a thick coating layer, and metal contacts. The second part $\partial \Lambda_A$ is only nonempty for devices embedded in integrated circuits and consists of boundary segments artificially introduced to separate the device from adjacent devices. The corresponding physical and artificial semiconductor and oxide boundary segments are denoted by $\partial \Omega_P$, $\partial \Omega_A$ and $\partial \Phi_P$, $\partial \Phi_A$ resp.
As an example we show an idealized MOS-transistor in Fig. 2.3.1 and a cross-section, taken in the (x_1, x_2)-plane, in Fig. 2.3.2. The transistor is embedded in an integrated circuit. The bulk, drain, source and gate contacts represented by the line segments $\overline{HG}$, $\overline{EF}$, $\overline{AB}$ and $\overline{CD}$ resp. in Fig. 2.3.2 are 'physical' boundaries and the other segments are 'artificial'.
Artificial boundaries have to be introduced such that they do not significantly perturb the model. Therefore a-priori information on the device has to be used.
Artificial boundaries are sometimes also introduced to simplify the numerical simulation by 'cutting off' regions of the device with insignificant action (see [2.16]).
In order to make the device self-contained a vanishing outward electric field

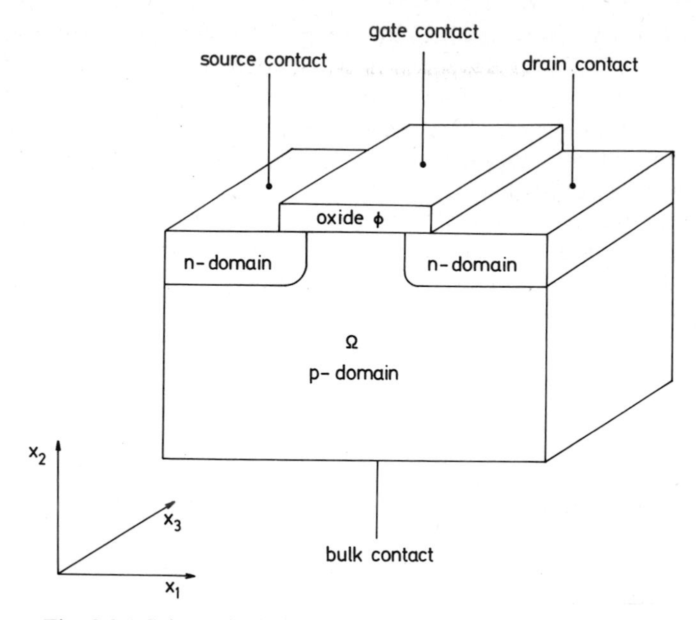

Fig. 2.3.1 Schematic diagram of an N-channel MOS-transistor

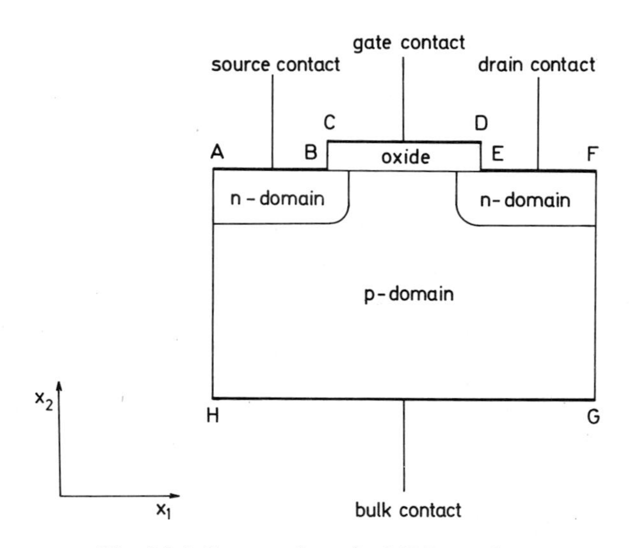

Fig. 2.3.2 Cross-section of a MOS-transistor

component on the artificial device boundary $\partial \Lambda_A$ and vanishing outward current density components on the artificial semiconductor domain-boundaries $\partial \Omega_A$ are usually assumed:

$$E \cdot v|_{\partial \Omega_A} = E \cdot \xi|_{\partial \Phi_A} = J_n \cdot v|_{\partial \Omega_A} = J_p \cdot v|_{\partial \Omega_A} = 0, \qquad (2.3.1)$$

where we denote by ν the exterior unit normal vector of the semiconductor boundary $\partial\Omega$ and by ξ the exterior unit normal vector of the oxide boundary $\partial\Phi$.
The 'physical' semiconductor boundary part $\partial\Omega_P$ generally splits into a finite number of contact segments, insulating segments and semiconductor-oxide interfaces (for a MOS-device).
Metal semiconductor contacts are fabricated by bringing a metal into intimate contact with the semiconductor. Mathematically, a contact is represented by a closed and connected subset of $\partial\Omega_P$.
Generally, a relationship between the potential, its time derivative, the outflow current, its time derivative, the applied bias and the 'forced' current is prescribed at the contact. For a pure voltage drive, the potential at the contact as function of time is prescribed and for a pure current drive the outflow current as function of time is given.
In some applications, mixed current-voltage driven contacts occur. We shall, however, in the following mainly consider purely voltage driven contacts, and, in some case, purely current driven contacts.
Two different types of semiconductor contacts are used in modern device technology, namely Ohmic and Schottky contacts.
An Ohmic contact, denoted by O in the sequel, has a negligible contact resistance relative to the bulk or spreading resistance of the semiconductor (see [2.20]). It does not significantly perturb the device performance. Usually, thermal equilibrium

$$np|_O = n_i^2|_O \tag{2.3.2}$$

and a vanishing space charge

$$(n-p-C)|_O = 0 \tag{2.3.3}$$

is assumed to hold at Ohmic contacts. Dirichlet boundary conditions for n and p can easily be obtained from (2.3.2), (2.3.3):

$$n|_O = \frac{1}{2}(C+\sqrt{C^2+4n_i^2})|_O, \quad p|_O = \frac{1}{2}(-C+\sqrt{C^2+4n_i^2})|_O. \tag{2.3.4}$$

The boundary potential at an Ohmic contact is the sum of the potential $U_O(t)$, which is externally applied to the contact O, and the so called built-in-potential, which is produced by the doping:

$$\psi|_O = \psi_{\mathrm{bi}}|_O + U_O(t). \tag{2.3.5}$$

The built-in potential ψ_{bi} is chosen such that the device is in thermal equilibrium, if all externally applied potentials are zero. Since the current densities J_n and J_p vanish in thermal equilibrium, we obtain by integrating (2.1.35), (2.1.36):

$$n_e = Ae^{\frac{\psi_e}{U_T}}, \quad p_e = Be^{-\frac{\psi_e}{U_T}}, \tag{2.3.6}$$

where ψ_e denotes the equilibrium potential. The mass-action law (2.2.1) gives:

$$AB = n_i^2. \tag{2.3.7}$$

Since an arbitrary constant can be added to the potential without changing the carrier concentrations, we choose $A = B = n_i$ and obtain:

$$n_e = n_i e^{\frac{\psi_e}{U_T}}, \quad p_e = n_i e^{-\frac{\psi_e}{U_T}}. \tag{2.3.8}$$

Using (2.3.3) we calculate the equilibrium potential at the Ohmic contact $\psi_e|_O$, which equals $\psi_{\mathrm{bi}}|_O$, and obtain by extending the built-in-potential to Ω:

$$\psi_{\mathrm{bi}}(x) = U_T \ln\left[\frac{C(x)+\sqrt{C^2(x)+4n_i^2}}{2n_i}\right]. \tag{2.3.9}$$

The boundary data for n and p are independent of time, the boundary datum for the potential is time dependent, if and only if the externally applied potential is time dependent.
Examples for Ohmic contacts are provided by the source, drain and bulk contacts of the MOS-transistor of Fig. 2.3.1.
A Schottky contact is a metal-semiconductor contact with a significantly large metal-semiconductor barrier height. It influences the device performance by generating a thin region of large space charge (see [2.20]). The physics of Schottky contacts is not fully understood yet and the models used for simulation are highly simplifying.
For a purely voltage driven Schottky contact, denoted by S in the sequel, the following Dirichlet condition for the potential is usually employed:

$$\psi|_S = \psi_{\mathrm{bi}}|_S - \varphi_S + U_S(t), \tag{2.3.10}$$

where φ_S denotes the metal-semiconductor (Schottky) barrier height. Typical numerical values for φ_S are between half a Volt and one Volt at room temperature (see [2.20]).
The boundary values for n and p at a Schottky contact depend on the current densities at the contact. The following conditions are derived from the thermionic emission and diffusion theories (see [2.20]):

$$J_n \cdot v|_S = -qv_n\left(n - \frac{C+\sqrt{C^2+4n_i^2}}{2}e^{-\varphi_S}\right)\Bigg|_S \tag{2.3.11}$$

$$J_p \cdot v|_S = qv_p\left(p - \frac{-C+\sqrt{C^2+4n_i^2}}{2}e^{\varphi_S}\right)\Bigg|_S. \tag{2.3.12}$$

v_n and v_p denote the thermionic recombination velocities for electrons and holes at the contacts (see [2.20] for numerical values).
The insulating semiconductor boundary part $\partial\Omega_{\mathrm{ins}}$, mathematically represented by a union of finitely many open segments, is assumed to be ideal, i.g. we neglect interface charges and surface recombination and assume that the electric field in the thick insulating layer vanishes. Then the outward electric field and the outward current density components at $\partial\Omega_{\mathrm{ins}}$ vanish:

$$E \cdot v|_{\partial\Omega_{\mathrm{ins}}} = J_n \cdot v|_{\partial\Omega_{\mathrm{ins}}} = J_p \cdot v|_{\partial\Omega_{\mathrm{ins}}} = 0. \tag{2.3.13}$$

At the insulating oxide-boundaries we assume:

$$E \cdot \xi \big|_{\partial \Phi_{\text{ins}}} = 0. \tag{2.3.14}$$

The semiconductor-oxide interface $\partial\Omega_I$, which is non-empty for MOS-devices only, is the interior with respect to $\partial\Omega$ of the joint boundary segments of the semiconductor domain Ω and the union of oxide domains Φ, i.e. $\overline{\partial\Omega_I} = \overline{\partial\Omega \cup \partial\Phi}$, where '$-$' denotes the closure.
The interface of the MOS-transitor depicted in Fig. 2.3.1 is the plane segment 'between' the semiconductor and oxide cuboids. In the cross-sectional view (Fig. 2.3.2) it is represented by the line segment $\overline{BE}$.
Neglecting interface charges, we assume that the potential and the electrical displacement components in direction orthogonal to the interface are continuous:

$$[\psi]_{\partial\Omega_I} = [D \cdot v]_{\partial\Omega_I} = 0. \tag{2.3.15}$$

Here $[f]_{\partial\Omega_I}$ denotes the height of the 'jump' of the function f across the surface $\partial\Omega_I$. Assuming that the permittivity is piecewise constant, i.e.

$$\varepsilon(x) = \begin{cases} \varepsilon_o, & x \in \Phi \\ \varepsilon_s, & x \in \Omega \end{cases},$$

where ε_s and ε_o denote the semiconductor and oxide permittivities resp. (see Table 2.1.1 for numerical value), we obtain by using (2.1.4) (a):

$$[\varepsilon E \cdot v]_{\partial\Omega_I} = 0 \tag{2.3.16}$$

We neglect the existence of mobile carriers in the oxide, thus $J_n|_\Phi \equiv \equiv J_p|_\Phi \equiv 0$. Also we neglect surface recombination effects on $\partial\Omega_I$ and obtain the boundary conditions:

$$J_n \cdot v \big|_{\partial\Omega_I} = J_p \cdot v \big|_{\partial\Omega_I} = 0. \tag{2.3.17}$$

No current is assumed to flow through the interface.
In practice every oxide domain has a metal contact (gate contact). The boundary condition for the potential at the gate contact G reads:

$$\psi_G = -U_F - \varphi_G + U_G(t), \tag{2.3.18}$$

where U_G denotes the externally applied gate potential, φ_G the metal-semiconductor work function difference and U_F the flat band voltage (see [2.20]). Typically $\varphi_G \approx -0.6$ V and $U_F \approx 1$ V for a silicon-silicon dioxide device at room temperature.
From a mathematical point of view, there is no need to distinguish between insulating segments and artificial boundary segments, since the same boundary conditions hold there. For the following we therefore split $\partial\Omega$ into the three disjoint subsets:

$$\partial\Omega = \partial\Omega_D \cup \partial\Omega_N \cup \partial\Omega_I, \tag{2.3.19}$$

where $\partial\Omega_N = \partial\Omega_A \cup \partial\Omega_{\text{ins}}$. $\partial\Omega_D$ denotes the union of contact segments, which splits into the union of Ohmic and the union of Schottky contacts. Analogously we split $\partial\Phi$ into

$$\partial\Phi = \partial\Phi_N \cup \partial\Phi_D \cup \overline{\partial\Omega_I} \tag{2.3.20}$$

with $\partial\Phi_N = \partial\Phi_A \cup \partial\Phi_{\text{ins}}$. $\partial\Phi_D$ denotes the union of oxide contacts.
We now rewrite the boundary conditions using $E = -\operatorname{grad}\psi$ and the splitting (2.3.19), (2.3.20):

$$\left.\frac{\partial\psi}{\partial\nu}\right|_{\partial\Omega_N} = \left.\frac{\partial\psi}{\partial\xi}\right|_{\partial\Phi_N} = J_n \cdot \nu|_{\partial\Omega_N \cup \partial\Omega_I} = J_p \cdot \nu|_{\partial\Omega_N \cup \partial\Omega_I} = 0, \tag{2.3.21}$$

$$[\psi]_{\partial\Omega_I} = \left[\varepsilon\frac{\partial\psi}{\partial\nu}\right]_{\partial\Omega_I} = 0. \tag{2.3.22}$$

The Static Problem

In order to completely define the semiconductor device problem, we have to prescribe the carrier concentrations at, say, time $t = 0$:

$$n(x, t = 0) = n_0(x), \quad p(x, t = 0) = p_0(x), \quad x \in \Omega. \tag{2.3.23}$$

If all potentials, which are externally applied to the contacts, are time independent, then the device problem is reduced to a stationary problem by ignoring the initial conditions and by assuming that ψ, n, p, J_n, J_p are time-independent:

$$\psi = \psi(x),\ n = n(x),\ p = p(x),\ J_n = J_n(x),\ J_p = J_p(x). \tag{2.3.24}$$

Consequently, we set

$$\frac{\partial n}{\partial t} = \frac{\partial p}{\partial t} = 0 \tag{2.3.25}$$

in the continuity equations (2.1.33), (2.1.34).
Steady state solutions can very often be interpreted as limits (as time tends to infinity) of solutions of the transient problem with (asymptotically) equal boundary conditions and arbitrary initial data. Thus the main goal of static simulation is the investigation of the 'large-time' performance of semiconductor devices, i.e. of the possible states sufficiently long after switching processes.

Geometric Properties of Devices

Semiconductor devices are three-dimensional structures. However, many devices are 'intrinsically two or even one-dimensional' because of their special geometries and doping profiles. An example is provided by the schematic diagram of the MOS-transistor depicted in Fig. 2.3.1. In practice, the doping is homogeneous in x_3-direction, i.e. $C(x) = C(x_1, x_2)$, over a very large region of the device. Since the parameters R, μ_n, μ_p and the boundary data only depend on ψ, n, p, J_n, J_p and on the doping profile C, but not on the

position vector explicitly, we assume – facilitated by the location of the contacts as depicted in Fig. 2.3.1 – that the solutions ψ, n, p, J_n, J_p of the MOS-transistor device problem are independent of x_3. Then the current density components in x_3-direction vanish. Therefore, the device problem is reduced to two space dimensions by cancelling out the derivatives with respect to x_3 and the x_3-components of J_n and J_p. Usually, the error introduced by this simplification is reasonably small.
The two-dimensional simulation geometry is a cross section of the device domain parallel to the (x_1, x_2)-plane. As depicted in Fig. 2.3.2, it consists of two adjacent rectangles.
In the following we shall not distinguish between the physical, three-dimensional device domain and the one, two or three-dimensional simulation domain, denote both by Λ and, for $\Lambda \subseteq \mathbb{R}^k$ refer to the problem as k-dimensional semiconductor device problem. Also we shall denote by Ω and Φ those k-dimensional subdomains of $\Lambda \subseteq \mathbb{R}^k$, which correspond to the physical semiconductor and oxide domains resp.
Some semiconductor devices have a cylindrically symmetric geometry and doping profile. Numerical simulations of such devices were successfully performed by transforming the problem to cylindrical coordinates (see [2.6]), thus reducing it by one space dimension.

2.4 Dependent Variables and Scaling – Transformation of State Variables

When Boltzmann's statistics are assumed to hold, then the carrier concentrations can be written as (see [2.16]):

$$n = n_i \exp\left(\frac{\psi - \varphi_n}{U_T}\right), \qquad p = n_i \exp\left(\frac{\varphi_p - \psi}{U_T}\right), \tag{2.4.1}$$

where φ_n and φ_p denote the electron and hole quasi-Fermi potentials. From a mathematical point of view we can regard (2.4.1) as change of dependent variables. $(\psi, n, p) \in \mathbb{R} \times (0, \infty)^2$ is one-to-one related to $(\psi, \varphi_n, \varphi_p) \in \mathbb{R}^3$. The validity of Boltzmann's statistics is only necessary to allow the physical interpretation of φ_n and φ_p as quasi-Fermi potentials.
The current relations (2.1.35), (2.1.36) are easily expressed in terms of $(\psi, \varphi_n, \varphi_p)$:

$$J_n = -q\mu_n n_i \exp\left(\frac{\psi - \varphi_n}{U_T}\right) \operatorname{grad} \varphi_n = -q\mu_n n \operatorname{grad} \varphi_n, \tag{2.4.2 (a)}$$

$$J_p = -q\mu_p n_i \exp\left(\frac{\varphi_p - \psi}{U_T}\right) \operatorname{grad} \varphi_p = -q\mu_p p \operatorname{grad} \varphi_p. \tag{2.4.2 (b)}$$

Another set of variables is obtained by setting:

$$u = \exp\left(-\frac{\varphi_n}{U_T}\right), \qquad v = \exp\left(\frac{\varphi_p}{U_T}\right), \tag{2.4.3}$$

and, by using (2.4.1):

$$n = n_i e^{\frac{\psi}{U_T}} u, \qquad p = n_i e^{-\frac{\psi}{U_T}} v. \tag{2.4.4}$$

Then the current relations read:

$$J_n = q\mu_n U_T n_i e^{\frac{\psi}{U_T}} \operatorname{grad} u, \qquad J_p = -q\mu_p U_T n_i e^{-\frac{\psi}{U_T}} \operatorname{grad} v. \tag{2.4.5}$$

In thermal equilibrium $\varphi_n = \varphi_p = 0$ and correspondingly $u = v = 1$ holds. We remark that n and p must be positive because of their physical interpretation as carrier concentrations. This is properly reflected by the formulas (2.4.1), if, however, the set of variables (ψ, u, v) is used, then we additionally have to require $u > 0$, $v > 0$ in Ω.
We shall see later on that the set (ψ, n, p) is generally the best choice of dependent variables for numerical simulations, while the set (ψ, u, v), or equivalently $(\psi, \varphi_n, \varphi_p)$, simplifies and to a great extent facilitates the mathematical analysis of the stationary problem.

The Singular Perturbation Scaling

An important step towards the structural analysis of a mathematical model is an appropriate scaling, which introduces dimensionless quantities and isolates the relevant dimensionless parameters on which the model depends (see [2.11]).
The scaling we shall present here is given in the references [2.22], [2.23], [2.12], [2.13]. It is based on the following premises:

(i) Length is referred to a characteristic device length l chosen such that l is of the same order of magnitude as the diameter of the semiconductor simulation domain Ω:

$$l = O(\operatorname{diam}(\Omega)) \tag{2.4.6}$$

(ii) All potentials are scaled by the thermal voltage U_T.

(iii) The carrier concentrations are scaled by a characteristic doping concentration $\tilde{C}$, which is chosen such that the maximal doping concentration

$$C_{\max} := \max_{\Omega} |C(x)| \tag{2.4.7}$$

is of the same order of magnitude as $\tilde{C}$:

$$\tilde{C} = O(C_{\max}). \tag{2.4.8}$$

Then the scaled majority carrier concentrations at Ohmic contacts are maximally of the order of magnitude 1.

(iv) The mobilities are referred to a reference carrier mobility $\tilde{\mu}$ chosen such that $\mu_n/\tilde{\mu}$ and $\mu_p/\tilde{\mu}$ are maximally of the order of magnitude 1.

The scaling factors of the involved quantities together with typical numerical values are given in Table 2.4.1. The scaled quantity is the ratio of the unscaled physical quantity and the corresponding scaling factor.

Table 2.4.1. Scaling factors

Symbol	Meaning	Scaling factor	Typical numerical value
x	space variable	l	5×10^{-3} cm
t	time variable	$\frac{l^2}{\tilde{\mu} U_T}$	9.7×10^{-7} s
ψ	potential	U_T	0.0259 V
n	electron concentration	$\tilde{C}$	10^{17} cm^{-3}
p	hole concentration	$\tilde{C}$	10^{17} cm^{-3}
J_n	electron current density	$\frac{qU_T\tilde{C}\tilde{\mu}}{l}$	83 Acm^{-2}
J_p	hole current density	$\frac{qU_T C\tilde{\mu}}{l}$	83 Acm^{-2}
μ_n	electron mobility	$\tilde{\mu}$	1000 cm^2 V^{-1} s^{-1}
μ_p	hole mobility	$\tilde{\mu}$	1000 cm^2 V^{-1} s^{-1}
C	doping profile	$\tilde{C}$	10^{17} cm^{-3}
R	recombination-generation rate	$\frac{U_T\tilde{\mu}\tilde{C}}{l^2}$	1.1×10^{23} cm^{-3} s^{-1}
τ_n^l	electron lifetime	$\frac{l^2}{\tilde{\mu} U_T}$	9.7×10^{-7} s
τ_p^l	hole lifetime	$\frac{l^2}{\tilde{\mu} U_T}$	9.7×10^{-7} s
n_i	intrinsic carrier concentration	$\tilde{C}$	10^{17} cm^{-3}

We rewrite the semiconductor device equations (2.1.32)–(2.1.37) in terms of the scaled variables and obtain – denoting the scaled quantities by the same symbols as the corresponding unscaled quantities:

$$\lambda^2 \Delta\psi = n - p - C(x) \tag{2.4.9}$$

$$\operatorname{div} J_n - \frac{\partial n}{\partial t} = R \tag{2.4.10}$$

$$\operatorname{div} J_p + \frac{\partial p}{\partial t} = -R \tag{2.4.11}$$

$$J_n = \mu_n(\operatorname{grad} n - n \operatorname{grad} \psi) \tag{2.4.12}$$

$$J_p = -\mu_p(\operatorname{grad} p + p \operatorname{grad} \psi). \tag{2.4.13}$$

The scaled static problem is obtained by setting $\frac{\partial n}{\partial t} = \frac{\partial p}{\partial t} = 0$.

The system (2.4.9)–(2.4.13) holds in the 'scaled' semiconductor simulation domain, which we denote by Ω (as the unscaled domain). Also, we retain the symbol Φ for the 'scaled' oxide domain.

The parameter λ, whose square appears as multiplier of the Laplacian of the potential in the scaled Poisson's equation (2.4.9), is given by:

$$\lambda = \frac{\lambda_D}{l}, \qquad \lambda_D = \sqrt{\frac{\varepsilon_s U_T}{q\tilde{C}}}. \tag{2.4.14}$$

λ_D is the characteristic Debye length of the device under consideration (see [2.20]), thus λ is the normed characteristic Debye length of the device. For practically relevant devices λ is very small, typically of the order of magnitude $10^{-3} - 10^{-5}$.

Clearly, the scaled parameter functions R, μ_n, μ_p have to be expressed in terms of the scaled variables. For example, the scaled Shockley-Read-Hall recombination-generation term (see [2.2.2]) reads:

$$R_{\text{SRH}} = \frac{np - \delta^4}{\tau_p^l(n+\delta^2) + \tau_n^l(p+\delta^2)}. \tag{2.4.15}$$

δ^2 denotes the scaled intrinsic number:

$$\delta^2 = \frac{n_i}{\tilde{C}} \tag{2.4.16}$$

Typically, δ^2 is between 10^{-7} and 10^{-10}.

Also the boundary and initial conditons have to be expressed in terms of the scaled variables. The homogeneous conditions (2.3.21), (2.3.22) for the Neumann segments $\partial\Omega_N$ and the semiconductor-oxide interface $\partial\Omega_I$ remain unchanged. The scaled boundary conditions for an Ohmic contact read (see (2.3.4), (2.3.5)):

$$n|_O = \frac{1}{2}(C + \sqrt{C^2 + 4\delta^4})|_O, \qquad p|_O = \frac{1}{2}(-C + \sqrt{C^2 + 4\delta^4})|_O \tag{2.4.17}$$

$$\psi|_O = \psi_{\text{bi}}|_O + V_O, \tag{2.4.18}$$

where the scaled built-in-potential is given by:

$$\psi_{\text{bi}}(x) = \ln\left[\frac{C(x) + \sqrt{C^2(x) + 4\delta^4}}{2\delta^2}\right]. \tag{2.4.19}$$

$V_O = U_O/U_T$ denotes the scaled applied contact potential.

In the sequel we shall often use the set of variables (ψ, u, v). The scaled transformation (2.4.4) reads:

$$n = \delta^2 e^{\psi} u, \qquad p = \delta^2 e^{-\psi} v \tag{2.4.20}$$

and the current relations transform to:

$$J_n = \delta^2 \mu_n e^{\psi} \operatorname{grad} u, \qquad J_p = -\delta^2 \mu_p e^{-\psi} \operatorname{grad} v. \tag{2.4.21}$$

The parameters λ and δ are determined by the doping, length, material and temperature of the device under consideration. Both parameters are small (compared to 1) in practical situations, but their impact on the solutions is entirely different. λ^2 multiplies the highest order derivatives of the potential and therefore we expect it to directly influence the variation of the solutions. This conjecture will be totally confirmed later on. Here we only remark that a problem, in which a small parameter multiplies a derivative of highest order, is – in the mathematical literature – called a singular perturbation problem.
The small parameter δ mainly determines the magnitude of the built-in potential.
Different ways of scaling can be found in the literature (see [2.3], [2.4], [2.14]). Of course, they also serve the purpose of introducing dimensionless quantities, but they do not isolate structure-parameters and they are by no means as appropriate for analytical and numerical purposes as the presented 'singular perturbation scaling'.

References

[2.1] Antoniadis, D. A., Hauser, S. E., Dutton, R. W.: SUPREM II: A Program for IC Process Modeling and Simulation. Report 5019–2, Stanford University, Cal., U.S.A., 1978.
[2.2] Antoniadis, D. A., Dutton, R. W.: Models for Computer Simulation of Complete IC Fabrication Processes. IEEE J. Solid State Circuits *SC 14,* No. 2, 412–422 (1979).
[2.3] De Mari, A.: An Accurate Numerical Steady State One-Dimensional Solution of the *P–N* Junction. Solid State Electron., *11,* 33–58 (1968).
[2.4] De Mari, A.: An Accurate Numerical One-Dimensional Solution of the *P–N* Junction under Arbitrary Transient Conditions. Solid State Electron. *11,* 1021–2053 (1968).
[2.5] Furikawa, S., Matsumura, H., Ishiwara, H.: Theoretical Considerations on Lateral Spread of Implanted Ions. Jap. J. Appl. Phys. *11,* No. 2, 134–142 (1972).
[2.6] Franz, G. A., Franz, A. F., Selberherr, S.: Cylindrically Symmetric Semiconductor Device Simulation. Report, Institut für Allgemeine Elektrotechnik, Technische Universität Wien, Austria, 1983.
[2.7] Hall, R. N.: Electron-Hole Recombination in Germanium. Physical Review *87,* 387 (1952).
[2.8] Hofmann, H.: Das elektromagnetische Feld, 2. Aufl. Wien–New York: Springer 1982.
[2.9] Jüngling, W.: Hochdotierungseffekte in Silizium. Diplomarbeit, Technische Universität Wien, Austria, 1983.
[2.10] Jüngling, W., Guerrero, E., Selberherr, S.: On Modeling the Intrinsic Number and Fermi Levels for Device and Process Simulation. COMPEL *3,* No. 2., 79–105 (1984).
[2.11] Lin, C. C., Segel, L. A.: Mathematics Applied to Deterministic Problems in the Natural Sciences. New York: Macmillan 1974.
[2.12] Markowich, P. A.: A Qualitative Analysis of the Fundamental Semiconductor Device Equations. COMPEL *2,* No. 3, 97–115 (1983).
[2.13] Markowich, P. A., Ringhofer, C.: A Singularly Perturbed Boundary Value Problem Modeling a Semiconductor Device. SIAM J. Appl. Math. *44,* No. 2, 231–256 (1984).
[2.14] Mock, M. S.: Analysis of Mathematical Models of Semiconductor Devices. Dublin: Boole Press 1983.
[2.15] Schütz, A.: Simulation des Lawinendurchbruchs in MOS-Transistoren. Dissertation, Technische Universität Wien, Austria, 1982.

[2.16] Selberherr, S.: Analysis and Simulation of Semiconductor Devices. Wien–New York: Springer 1984.
[2.17] Selberherr, S., Griebel, W., Pötzl, H.: Transport Physics for Modelling Semiconductor Devices, Proceedings, Conference "Simulation of Semiconductor Devices", Swansea, 1984.
[2.18] Shockley, W., Read, W. T.: Statistics of the Recombination of Holes and Electrons. Physical Review, *87,* No. 5, 835–842 (1952).
[2.19] Smith, R. A.: Semiconductor, 2nd ed. Cambridge: Cambridge University Press 1978.
[2.20] Sze, S. M.: Physics of Semiconductor Devides, 2nd ed. New York: J. Wiley 1981.
[2.21] Van Roosbroeck, W. V.: Theory of Flow of Electrons and Holes in Germanium and Other Semiconductors. Bell Syst. Tech. J. *29,* 560–607 (1950).
[2.22] Vasileva, A. B., Butuzow, V. F.: Singularly Perturbed Equations in the Critical Case. MRC–TSR 2039, Math. Res. Center, University Wisconsin-Madison, U.S.A., 1980.
[2.23] Vasileva, A. B., Stelmakh, V. G.: Singularly Disturbed Systems of the Theory of Semiconductor Devices. USSR Comput. Math. Phys. *17,* 48–58 (1977).
[2.24] Zimam, J. M.: Electrons and Phonons. London: Clarendon Press 1963.

Analysis of the Basic Stationary Semiconductor Device Equations 3

3.1 Preliminaries

In this chapter we present existence, regularity, uniqueness and continuous-dependence-on-data results for the basic semiconductor device equations in the stationary case. The analysis will heavily rely on the modern abstract theory of elliptic differential equations and on functional analysis. The results on elliptic boundary value problems, which we shall use in the sequel, can be found in the references [3.7], [3.10], [3.14], [3.20], [3.32], the concepts from functional analysis in [3.5], [3.6], [3.10], [3.13], [3.21], [3.32] and the theory of Sobolev spaces in [3.1]. A list of notations is compiled in the Appendix. Although the scaling does not influence the results to be presented in this chapter we shall work with the singular perturbation formulation of the stationary basic semiconductor device equations as derived in Section 2.4:

$$\lambda^2 \Delta\psi = n-p-C(x) \qquad \text{(Poisson's equation)} \tag{3.1.1}$$

$$\operatorname{div} J_n = R \qquad \text{(electron continuity equation)} \tag{3.1.2}$$

$$\operatorname{div} J_p = -R \qquad \text{(hole continuity equation)} \tag{3.1.3}$$

$$J_n = \mu_n(\operatorname{grad} n - n \operatorname{grad} \psi) \qquad \text{(electron current relation)} \tag{3.1.4}$$

$$J_p = -\mu_p(\operatorname{grad} p + p \operatorname{grad} \psi) \qquad \text{(hole current relation)} \tag{3.1.5}$$

The equations (3.1.1)–(3.1.5) are posed in a bounded domain $\Omega \subseteq \mathbb{R}^k$, $k = 1, 2$ or 3, representing the semiconductor part of the device geometry. The interaction of the device under consideration with the outer world is described by boundary conditions. In the most general set-up they are discussed in Section 2.3, we shall however assume in this chapter that the boundary $\partial\Omega$ splits into 'Neumann segments' $\partial\Omega_N$ and 'Dirichlet segments' $\partial\Omega_D$ only:

$$\partial\Omega = \partial\Omega_N \cup \partial\Omega_D, \qquad \partial\Omega_N \cap \partial\Omega_D = \{\ \}. \tag{3.1.6}$$

$\partial\Omega_N$ represents the union of artificial and insulating boundary segments. The electric field $E = -\operatorname{grad}\psi$ in outward direction and the outward current density components vanish there:

$$E \cdot \nu|_{\partial\Omega_N} = J_n \cdot \nu|_{\partial\Omega_N} = J_p \cdot \nu|_{\partial\Omega_N} = 0. \tag{3.1.7}$$

ν denotes the outward unit normal vector of $\partial\Omega$.
Dirichlet boundary conditions are assumed to hold on $\partial\Omega_D$:

$$\psi|_{\partial\Omega_D} = \psi_D|_{\partial\Omega_D}, \quad n|_{\partial\Omega_D} = n_D|_{\partial\Omega_D}, \quad p|_{\partial\Omega_D} = p_D|_{\partial\Omega_D}. \tag{3.1.8}$$

For analytical convenience we take the Dirichlet data to be traces of functions ψ_D, n_D, p_D, which are defined on $\bar{\Omega}$.
Physically, $\partial\Omega_D$ is the union of, say, r Ohmic contacts O_i:

$$\partial\Omega_D = \bigcup_{i=1}^{r} O_i. \tag{3.1.9}$$

From Section 2.4 we obtain:

$$\begin{aligned} n_D|_{\partial\Omega_D} &= \frac{1}{2}\left(C + \sqrt{C^2 + 4\delta^4}\right)|_{\partial\Omega_D}, \\ p_D|_{\partial\Omega_D} &= \frac{1}{2}\left(-C + \sqrt{C^2 + 4\delta^4}\right)|_{\partial\Omega_D}, \end{aligned} \tag{3.1.10 (a)}$$

$$\begin{aligned} \psi_D|_{O_i} &= \psi_{\mathrm{bi}}|_{O_i} + U_i, \quad i = 1, \ldots, r, \\ \psi_{\mathrm{bi}}(x) &:= \ln\left[\frac{C(x) + \sqrt{C^2(x) + 4\delta^4}}{2\delta^2}\right]. \end{aligned} \tag{3.1.10 (b)}$$

U_i denotes the scaled static potential, which is externally applied to $O_i \cdot \partial\Omega_N$ is open with respect to $\partial\Omega$ and $\partial\Omega_D$ is closed. The O_i's are closed, connected and pairwise disjoint.
We do not treat MOS-devices and devices with Schottky contacts in this chapter. Most of the results presented in the sequel, in particular the existence theorem, can easily be extended by minor modifications of the proofs.
A device is specified when its material, its geometry and the doping profile $C(x)$ are given. Physically, the operating temperature and the static externally applied potentials then determine (not necessarily in a unique way) the static potential, carrier concentrations and current densities in the device. In the next sections we shall show that the static semiconductor device problem is 'well-posed' in a certain mathematical sense, which reflects the most important physical properties of semiconductor devices like existence of stationary states, their continuous dependence on data and their uniqueness under near-equilibrium conditions. These results are fundamentally important for the subsequent singular perturbation analysis and the discussion of discretisation techniques.
Analytical results on the stationary basic semiconductor device equations were obtained by many authors using various assumptions on the mobilities, the recombination-generation rate and the geometry (see [3.17], [3.23], [3.9], [3.15], [3.3], [3.11a]; [3.19] gives a rather comprehensive survey). We shall collect and extend these results to allow for physically appropriate parameter models and geometries.

3.2. Existence of Solutions

In this section we address the question of existence of solutions of the stationary semiconductor device problem. Therefore we insert the current relations (3.1.4), (3.1.5) into the continuity equations (3.1.2) and (3.1.3) resp. and obtain

$$\operatorname{div}(\mu_n(\operatorname{grad} n - n \operatorname{grad} \psi)) = R \tag{3.2.1}$$

$$\operatorname{div}(\mu_p(\operatorname{grad} p + p \operatorname{grad} \psi)) = R \tag{3.2.2}$$

For given potential ψ and given recombination-generation rate R both equations are elliptic since the mobilities μ_n, μ_p are positive. However, a direct analysis of (3.2.1), (3.2.2) is cumbersome, particularly because the drift terms $-n \operatorname{grad} \psi$ and $p \operatorname{grad} \psi$ render the application of a powerful tool for the analysis of elliptic partial differential equations, namely the maximum principle, inapplicable. The subsequent analysis is to a large extent facilitated by the introduction of a new set of variables (ψ, u, v), which relates to (ψ, n, p) in the following way:

$$n = \delta^2 e^{\psi} u, \qquad p = \delta^2 e^{-\psi} v \tag{3.2.3}$$

The current relations then reduce to

$$J_n = \delta^2 \mu_n e^{\psi} \operatorname{grad} u \tag{3.2.4 (a)}$$
$$J_p = -\delta^2 \mu_p e^{-\psi} \operatorname{grad} v \tag{3.2.4 (b)}$$

and the following system of second order differential equations is obtained from (3.1.1) (3.1.5):

$$\left.\begin{aligned} &(SD1)\quad \lambda^2 \Delta\psi = \delta^2 e^{\psi} u - \delta^2 e^{-\psi} v - C(x) \\ &(SD2)\quad \operatorname{div}(\mu_n e^{\psi} \operatorname{grad} u) = S \\ &(SD3)\quad \operatorname{div}(\mu_p e^{-\psi} \operatorname{grad} v) = S \end{aligned}\right\} x \in \Omega.$$

The term S denotes the modified recombination-generation rate:

$$S = \frac{R}{\delta^2}. \tag{3.2.5}$$

The continuity equations $(SD2)$, $(SD3)$ are – for given S – selfadjoint. The boundary conditions (3.1.7) on $\partial\Omega_N$ transform to homogeneous Neumann conditions when expressed in (ψ, u, v):

$$(SD4)\quad \left.\frac{\partial \psi}{\partial \nu}\right|_{\partial\Omega_N} = \left.\frac{\partial u}{\partial \nu}\right|_{\partial\Omega_N} = \left.\frac{\partial v}{\partial \nu}\right|_{\partial\Omega_N} = 0$$

and the Dirichlet boundary conditions (3.1.8) now read:

$$(SD5)\quad \psi|_{\partial\Omega_D} = \psi_D|_{\partial\Omega_D}, \quad u|_{\partial\Omega_D} = u_D|_{\partial\Omega_D}, \quad v|_{\partial\Omega_D} = v_D|_{\partial\Omega_D},$$

where $n_D = \delta^2 e^{\psi_D} u_D$, $p_D = \delta^2 e^{-\psi_D} v_D$.

For Ohmic contacts we obtain from (3.1.10):

$$u_D\big|_{O_i} = e^{-U_i}, \quad v_D\big|_{O_i} = e^{U_i}, \quad i = 1, \ldots, r. \tag{3.2.6}$$

Since n and p represent 'physical' concentrations, we have to require $n(x) > 0, p(x) > 0$ in Ω. Expressed in the new set of variables this means that we only admit solutions ψ, u, v of the semiconductor problem which satisfy:

$$u(x) > 0, \quad v(x) > 0 \text{ in } \Omega. \tag{3.2.7}$$

We now state the assumptions on the geometry, data and parameter models, which will be used for the subsequent existence proof.

(A.3.2.1) Ω is a bounded domain in $\mathbb{R}^k$ ($k = 1, 2$ or 3) of class $C^{0,1}$ and the $(k-1)$-dimensional Lebesguemeasure of $\partial\Omega_D$ is positive.

(A.3.2.2) The Dirichlet boundary data satisfy

$$(\psi_D, u_D, v_D) \in (H^1(\Omega))^3, \quad (\psi_D, u_D, v_D)\big|_{\partial\Omega_D} \in (L^\infty(\partial\Omega_D))^3$$

and there is $U_+ \geqq 0$ such that

$$e^{-U_+} \leqq \inf_{\partial\Omega_D} u_D, \ \inf_{\partial\Omega_D} v_D; \quad \sup_{\partial\Omega_D} u_D, \ \sup_{\partial\Omega_D} v_D \leqq e^{U_+}.$$

(A.3.2.3) The doping profile satisfies $C \in L^\infty(\Omega)$. We denote $\underline{C} := \inf_\Omega C(x)$, $\bar{C} := \sup_\Omega C(x)$.

(A.3.2.4) The recombination-generation rate S satisfies $S = F(x, \psi, u, v)\,(uv-1)$, where $F(x, ., ., .) \in C^1(\mathbb{R} \times (0, \infty)^2)$ for all $x \in \Omega$; $F(., \psi, u, v)$, $\text{grad}_{(\psi, u, v)} F(., \psi, u, v) \in L^\infty(\Omega)$ uniformly for (ψ, u, v) in bounded sets of $\mathbb{R} \times (0, \infty)^2$ and $F(x, \psi, u, v) \geqq 0$ in $\Omega \times \mathbb{R} \times (0, \infty)^2$.

(A.3.2.5) The mobilities μ_n, μ_p satisfy:

(i) $\mu_n = \mu_n(x, \text{grad}\,\psi)$, $\mu_p = \mu_p(x, \text{grad}\,\psi)$, $\mu_n, \mu_p \colon \Omega \times \mathbb{R}^k \to \mathbb{R}$

(ii) $0 < \underline{\mu}_n \leqq \mu_n \leqq \bar{\mu}_n$ in $\Omega \times \mathbb{R}^k$ for some $\underline{\mu}_n, \bar{\mu}_n \in \mathbb{R}$

$0 < \underline{\mu}_p \leqq \mu_p \leqq \bar{\mu}_p$ in $\Omega \times \mathbb{R}^k$ for some $\underline{\mu}_p, \bar{\mu}_p \in \mathbb{R}$

(iii) $|\mu_n(x, y_1) - \mu_n(x, y_2)| + |\mu_p(x, y_1) - \mu_p(x, y_2)| \leqq L|y_1 - y_2|$ for all $x \in \Omega$; $y_1, y_2 \in \mathbb{R}^k$.

Clearly, the assumptions on the geometry (A.3.2.1) are realistic. Physical device boundaries are smooth, and at least Lipschitz-continuous boundaries are used for simulations. Also, every physical device has contacts with nonvanishing 'areas'. (A.3.2.2) is necessary to guarantee L^∞-solutions with u and v uniformly bounded away from zero. Physical doping profiles are bounded, just as required by (A.3.2.3).
The assumptions (A.3.2.4) and (A.3.2.5) on the parameter models are more stringent.
(A.3.2.4) admits the Shockley-Read-Hall recombination-generation model and the Auger model but it excludes the avalanche-generation term modeling impact ionisation (see Section 2.2 for details on recombination-generation modeling). Physically, impact ionisation is negligible, if the externally applied potentials are not 'too' large. We shall later on discuss the effect of the avalanche generation term on the solutions for a simple model device.

(A.3.2.5) allows all commonly used inhomogeneous mobility models and some field-dependent models. It excludes current-density and carrier-concentration dependent models. We remark that the positive lower bounds $\underline{\mu}_n, \underline{\mu}_p$, which guarantee the uniform ellipticity of the continuity equations *(SD2), (SD3)*, do not admit the modeling of velocity saturation at high electric fields (see Section 2.2).
We now state the existence theorem for the problem *(SD1)–(SD5)* (referred to as problem *(SD)* in the sequel).

Theorem 3.2.1: *Let the assumptions (A.3.2.1)–(A.3.2.5) hold. Then the problem (SD) has a weak solution* $(\psi^*, u^*, v^*) \in (H^1(\Omega) \cap L^\infty(\Omega))^3$, *which satisfies the* L^∞*-estimates:*

$$e^{-U_+} \leqq u^*(x) \leqq e^{U_+} \quad a.e. \text{ in } \Omega \tag{3.2.8}$$

$$e^{-U_+} \leqq v^*(x) \leqq e^{U_+} \quad a.e. \text{ in } \Omega \tag{3.2.9}$$

$$\min\left(\inf_{\partial\Omega_D} \psi_D, \ln\left[\frac{\underline{C}+\sqrt{\underline{C}^2+4\delta^4}}{2\delta^2}\right] - U_+\right) \leqq \psi^*(x)$$

$$\leqq \max\left(\sup_{\partial\Omega_D} \psi_D, \ln\left[\frac{\bar{C}+\sqrt{\bar{C}^2+4\delta^4}}{2\delta^2}\right] + U_+\right) \quad a.e. \text{ in } \Omega. \tag{3.2.10}$$

The proof of this existence Theorem is based on decoupling the equations *(SD1), (SD2), (SD3)* and using a fixed point argument (Schauder's Theorem) thereafter. Note that the boundary conditions *(SD4), (SD5)* are decoupled already.
The fixed point operator is constructed as follows:

(A) For given $(u, v) = (u_0, v_0)$; $u_0, v_0 > 0$ we solve Poisson's equation

$$\lambda^2 \Delta \psi = \delta^2 e^{\psi} u_0 - \delta^2 e^{-\psi} v_0 - C(x),\ x \in \Omega \tag{3.2.11}$$

subject to the boundary conditions

$$\left.\frac{\partial \psi}{\partial \nu}\right|_{\partial\Omega_N} = 0, \qquad \psi|_{\partial\Omega_D} = \psi_D|_{\partial\Omega_D} \tag{3.2.12}$$

for $\psi = \psi_1$. (3.2.11) is a semilinear elliptic equation.

(B) We decouple the continuity equations *(SD2), (SD3)* from Poisson's equation by substituting ψ_1 for ψ in $\mu_n e^{\psi}$, $\mu_p e^{-\psi}$ and S. Then the so obtained equations are decoupled from each other by appropriately changing the arguments of S. We separately solve the linear elliptic boundary value problems:

$$\operatorname{div}(\mu_n(x, \operatorname{grad} \psi_1) e^{\psi_1} \operatorname{grad} u) = F(x, \psi_1, u_0, v_0)(u v_0 - 1) \tag{3.2.13}$$

$$\left.\frac{\partial u}{\partial \nu}\right|_{\partial\Omega_N} = 0, \qquad u|_{\partial\Omega_D} = u_D|_{\partial\Omega_D} \tag{3.2.14}$$

for $u = u_1$ and

$$\operatorname{div}(\mu_p(x, \operatorname{grad}\psi_1)e^{-\psi_1}\operatorname{grad} v) = F(x, \psi_1, u_0, v_0)(u_0 v - 1) \quad (3.2.15)$$

$$\left.\frac{\partial v}{\partial \nu}\right|_{\partial\Omega_N} = 0, \qquad v|_{\partial\Omega_D} = v_D|_{\partial\Omega_D} \quad (3.2.16)$$

for $v = v_1$.

We shall show below that the problems (3.2.11), (3.2.12); (3.2.13), (3.2.14) and (3.2.15), (3.2.16) are uniquely solvable, if (u_0, v_0) is in an appropriate subset of $(L^2(\Omega))^2$. For these (u_0, v_0) we define the operator H by $H(u_0, v_0) = (u_1, v_1)$. Clearly every fixed point (u^*, v^*) of H determines the weak solution (ψ^*, u^*, v^*) of the problem *(SD)*, where ψ^* is obtained by solving (3.2.11), (3.2.12) with u_0, v_0 substituted by u^*, v^*.
At first we show that Poisson's equation (3.2.11), (3.2.12) is uniquely soluble for appropriately given u_0, v_0, ψ_D.

Lemma 3.2.1: *Let $(u_0, v_0) \in (L^\infty(\Omega))^2$ satisfy $0 < \underline{u} \leqq u_0(x) \leqq \bar{u}$, $0 < \underline{v} \leqq v_0(x) \leqq \bar{v}$ a.e. in Ω and let $\psi_D \in H^1(\Omega)$, $\psi_D|_{\partial\Omega_D} \in L^\infty(\partial\Omega_D)$. Also assume that (A.3.2.1) and (A.3.2.3) hold. Then the problem (3.2.11), (3.2.12) has a unique weak solution $\psi = \psi_1 \in H^1(\Omega) \cap L^\infty(\Omega)$, which satisfies the estimate*

$$\min\left(\inf_{\partial\Omega_D}\psi_D, \ln\left[\frac{\underline{C}+\sqrt{\underline{C}^2+4\delta^4\bar{u}\underline{v}}}{2\delta^2\bar{u}}\right]\right) \leqq \psi_1(x)$$

$$\leqq \max\left(\sup_{\partial\Omega_D}\psi_D, \ln\left[\frac{\bar{C}+\sqrt{\bar{C}^2+4\delta^4\underline{u}\bar{v}}}{2\delta^2\underline{u}}\right]\right) \text{ a.e. in } \Omega. \quad (3.2.17)$$

Proof: We show that the bounds in (3.2.17) are upper and lower solutions resp. Therefore we estimate

$$\delta^2 e^{\psi}u_0 - \delta^2 e^{-\psi}v_0 - C(x) \geqq \delta^2\underline{u}e^{\psi} - \delta^2\bar{v}e^{-\psi} - \bar{C}$$

and

$$\delta^2 e^{\psi}u_0 - \delta^2 e^{-\psi}v_0 - C(x) \leqq \delta^2\bar{u}e^{\psi} - \delta^2\underline{v}e^{-\psi} - \underline{C}.$$

Thus, the solutions $\tilde{\psi}$ and $\underset{\sim}{\psi}$ of the algebraic equations

$$0 = \delta^2\underline{u}e^{\tilde{\psi}} - \delta^2\bar{v}e^{-\tilde{\psi}} - \bar{C}$$

$$0 = \delta^2\bar{u}e^{\underset{\sim}{\psi}} - \delta^2\underline{v}e^{-\underset{\sim}{\psi}} - \underline{C}$$

satisfy the inequalities

$$\lambda^2\Delta\tilde{\psi} - (\delta^2 e^{\tilde{\psi}}u_0 - \delta^2 e^{-\tilde{\psi}}v_0 - C(x)) \leqq 0$$

$$\lambda^2\Delta\underset{\sim}{\psi} - (\delta^2 e^{\underset{\sim}{\psi}}u_0 - \delta^2 e^{-\underset{\sim}{\psi}}v_0 - C(x)) \geqq 0.$$

(Note that $\tilde{\psi}$ and $\underset{\sim}{\psi}$ are constant!). The constants $\bar{\psi} := \max(\sup_{\partial\Omega_D}\psi_D, \tilde{\psi})$, $\underline{\psi} := \min(\inf_{\partial\Omega_D}\psi_D, \underset{\sim}{\psi})$ are upper and, resp., lower solutions of (3.2.11), (3.2.12) since the function

$$\xi(\psi, x) = \delta^2 e^{\psi}u_0(x) - \delta^2 e^{-\psi}v_0(x) - C(x)$$

is monotonically increasing in ψ. Explicit computation of $\bar{\psi}$ and ψ gives the estimate (3.2.17) for every weak solution $\psi_1 \in H^1(\Omega) \cap L^\infty(\Omega)$ (see [3.20], Section 9.2).

Assume now that there are two weak solutions $\psi_1, \psi_2 \in H^1(\Omega) \cap L^\infty(\Omega)$. By subtraction we obtain for $g = \psi_1 - \psi_2$:

$$\lambda^2 \Delta g = (\delta^2 e^{\xi_1(x)} u_0(x) + \delta^2 e^{-\xi_2(x)} v_0(x)) g$$

$$\left.\frac{\partial g}{\partial \nu}\right|_{\partial\Omega_N} = g|_{\partial\Omega_D} = 0$$

with $\xi_1(x), \xi_2(x)$ between $\psi_1(x)$ and $\psi_2(x)$. The maximum principle (see [3.10], [3.20]) immediately implies $g \equiv 0$ in Ω, thus there may at most be one weak solution.

Let $K > 0$. For $w \in L^2(\Omega)$ we define the cut-function

$$w_K(x) := \begin{cases} K, & w(x) \geqq K \\ w(x), & -K \leqq w(x) \leqq K. \\ -K, & w(x) \leqq -K \end{cases}$$

Clearly $w_K \in L^\infty(\Omega)$, and $w_K \in H^1(\Omega)$ if $w \in H^1(\Omega)$ (see [3.14], Section 2.3). We now choose $K = \max(|\bar{\psi}|, |\underline{\psi}|)$ and define the operator $M: L^2(\Omega) \times [0, 1] \to L^2(\Omega)$ by $M(y, \sigma) = z$, where z is the solution of

$$\lambda^2 \Delta z = \sigma(\delta^2 e^{y_K} u_0 - \delta^2 e^{-y_K} v_0 - C) \tag{3.2.18 (a)}$$

$$\left.\frac{\partial z}{\partial \nu}\right|_{\partial\Omega_N} = 0, \quad z|_{\partial\Omega_D} = \sigma \psi_D|_{\partial\Omega_D}. \tag{3.2.18 (b)}$$

Every fixed point ψ_1 of $M(., 1)$, which satisfies $|\psi_1(x)| \leqq K$ a.e. in Ω, is a weak solution of (3.2.11), (3.2.12). Let ψ_1 be some fixed point of $M(., 1)$. A standard reqularity result implies $\psi_1 \in C^1(\Omega)$. Thus the set $\Omega_+ \subseteq \Omega$ of points x at which $\psi_1(x) > K$ holds is open in Ω and the boundary of Ω_+ consists of points x at which either $\psi_1(x) = K$ or which are contained in $\partial\Omega$. Assume that Ω_+ is nonempty and let $x^* \in \Omega_+$. We denote by Ω_+^* the maximal connected component of Ω_+ containing x^*. Then $\psi_1|_{\Omega_+^*}$ solves

$$\lambda^2 \Delta \psi_1 = \delta^2 e^K u_0(x) - \delta^2 e^{-K} v_0(x) - C(x) \quad \text{in } \Omega_+^*$$

$$\left.\frac{\partial \psi_1}{\partial \nu}\right|_{\partial\Omega_+^* \cap \partial\Omega_N} = 0, \quad \psi_1|_{\partial\Omega_+^* \cap \partial\Omega_D} = \psi_D|_{\partial\Omega_+^* \cap \partial\Omega_D}, \quad \psi_1|_{\partial\Omega_+^* - \partial\Omega} = K.$$

Since $\delta^2 e^K u_0(x) - \delta^2 e^{-K} v_0(x) - C(x) \geqq 0$ in Ω and $\sup_{\partial\Omega_D} \psi_D \leqq K$ we conclude that $\bar{\psi}_1 = K$ is an upper solution of this problem.

Therefore Ω_+ is empty and $\psi_1(x) \leqq K$ in $\bar{\Omega}$. $\psi_1(x) \geqq -K$ in $\bar{\Omega}$ follows analogously.

Since the operator $y \to y_K$, $L^2(\Omega) \to L^2(\Omega)$ is continuous (see [3.14], Section 2.3), it is easily shown that the right hand side of (3.2.18) (a) depends continuously in $L^2(\Omega)$ on $(y, \sigma) \in L^2(\Omega) x [0, 1]$. Thus, by the continuous dependence of solutions of elliptic equations in $H^1(\Omega)$ on $L^2(\Omega)$-right hand

sides and $H^1(\Omega)$-boundary data (see [3.10], Chapter 8), continuity of M follows.
Also the range of M is bounded in $H^1(\Omega)$:

$$\|z\|_{1,2,\Omega} \leqq \text{const}\, \|\sigma(\delta^2 e^{y_K} u_0 - \delta^2 e^{-y_K} v_0 - C)\|_{2,\Omega}$$

$$\leqq \text{const}\, (\mu_k(\Omega))^{\frac{1}{2}} (\delta^2 e^K (\bar{u} + \bar{v}) + \bar{C}).$$

Since $H^1(\Omega)$ is compactly imbedded in $L^2(\Omega)$ (see [3.1], Chapter 6) we conclude that M is completely continuous.
$M(y, 0) = 0$ holds for all $y \in L^2(\Omega)$.
The maximum principle implies that $z = M(y, \sigma)$ satisfies $\|z\|_{\infty,\Omega} \leqq$ const independently of $y \in L^2(\Omega)$ and $\sigma \in [0, 1]$ since

$$\|\sigma(\delta^2 e^{y_K} u_0 - \delta^2 e^{-y_K} v_0 - C)\|_{\infty,\Omega} \leqq \delta^2 e^K (\bar{u} + \bar{v}) + \bar{C}.$$

Thus the Leray-Schauder Theorem (see [3.10]) proves the existence of a fixed point ψ_1 of $M(.,1)$. □

The unique solvability of the linear problems (3.2.13), (3.2.14) and (3.2.15), (3.2.16) is immediate:

Lemma 3.2.2: *Let the assumptions (A.3.2.1), (A.3.2.2), (A.3.2.4), (A.3.25) hold and assume that* $\psi_1 \in H^1(\Omega) \cap L^\infty(\Omega)$, $(u_0, v_0) \in (L^\infty(\Omega))^2$ *and* $e^{-U_+} \leqq$ $\leqq u_0(x), v_0(x) \leqq e^{U_+}$ *a.e. in* Ω *with* U_+ *given in (A.3.2.2). Then the problems (3.2.13), (3.2.14) and (3.2.15), (3.2.16) have unique weak solutions* $u_1, v_1 \in H^1(\Omega) \cap L^\infty(\Omega)$ *which satisfy*

$$e^{-U_+} \leqq u_1(x), v_1(x) \leqq e^{U_+} \quad \text{a.e. in } \Omega.$$

Proof: The nonnegativity assumptions on F and on u_0, v_0 imply the existence of weak solutions $u_1, v_1 \in H^1(\Omega)$ (see [3.10], Chapter 8). From [3.14], Section 3.13, we conclude $u_1, v_1 \in L^\infty(\Omega)$. $v_0(x) \geqq e^{-U_+}$ implies

$$F(x, \psi_1, u_0, v_0)\,(e^{U_+} v_0(x) - 1) \geqq 0.$$

Since $u_1|_{\partial\Omega_D} \leqq e^{U_+}$ holds, $\bar{u} := e^{U_+}$ is an upper solution of (3.2.13), (3.2.14) and $u_1(x) \leqq e^{U_+}$ a.e. in Ω follows. The other bounds are established analogously. □

The preceeding Lemmas imply that H is well defined:

$$H: N \to N \tag{3.2.19 (a)}$$

where

$$N = \{(u,v) \in (L^2(\Omega))^2 \,|\, e^{-U_+} \leqq u(x), v(x) \leqq e^{U_+} \text{ a.e. in } \Omega\}. \tag{3.2.19 (b)}$$

Note that N is a closed and convex subset of $(L^2(\Omega))^2$.
We shall now show that $H: N \to N$ is completely continuous. When this result is established we conclude the existence of a fixed point of H in N from Schauder's Theorem (see [3.10]). The existence assertion of Theorem 3.2.1 then follows immediately, the estimates (3.2.8), (3.2.9) are implied by the

definition of the set N and the estimate (3.2.10) for the potential follows from (3.2.17) with $\underline{u} = \underline{v} = e^{-U_+}$ and $\bar{u} = \bar{v} = e^{U_+}$.

Lemma 3.2.3: *Let the assumptions (A.3.2.1)–(A.3.2.5) hold. Then H maps N into a precompact subset of $(L^2(\Omega))^2$.*

Proof: The continuous dependence of the solutions $u = u_1$ and $v = v_1$ on the data implies the estimates

$$\|u_1\|_{1,2,\Omega} \leqq L_1(\|F(.,\psi_1,u_0,v_0)\|_{2,\Omega} + \|u_D\|_{1,2,\Omega}) \qquad (3.2.20)\text{ (a)}$$

$$\|v_1\|_{1,2,\Omega} \leqq L_2(\|F(.,\psi_1,u_0,v_0)\|_{2,\Omega} + \|v_D\|_{1,2,\Omega}), \qquad (3.2.20)\text{ (b)}$$

where L_1, L_2 only depend on Ω, $\underline{\mu}_n$, $\bar{\mu}_n$, $\underline{\mu}_p$, $\bar{\mu}_p$, U_+, $\|\psi_1\|_{\infty,\Omega}$ and $\|F(.,\psi_1,u_0,v_0)\|_{\infty,\Omega}$ (see [3.10]). The assumptions (A.3.2.1)–(A.3.2.5) and the estimate (3.2.17) imply

$$\|u_1\|_{1,2,\Omega} + \|v_1\|_{1,2,\Omega} \leqq \text{const} \qquad (3.2.21)$$

for all $(u_0, v_0) \in N$ and the Rellich-Kondrachov imbedding Theorem (see [3.1]) assures that $H(N)$ is precompact in $(L^2(\Omega))^2$. □

The following Lemmas are simple consequences of the well-posedness of the elliptic boundary value problems (3.2.11), (3.2.12); (3.2.13), (3.2.14) and (3.2.15), (3.2.16) (see [3.14], Section 3.5.). We leave the proofs to the reader.

Lemma 3.2.4: *Let (A.3.2.1), (A.3.2.3) hold and assume that $\psi_D \in H^1(\Omega)$, $\psi_D|_{\partial\Omega_D} \subset L^\infty(\partial\Omega_D)$. Then the map $(u_0, v_0) \to \psi_1$, $N \subseteq (L^2(\Omega))^2 \to H^1(\Omega)$ is uniformly Lipschitz continuous.*

Lemma 3.2.5: *Assume that (A.3.2.1), (A.3.2.2), (A.3.2.4), (A.3.2.5) hold and let the sequences $(u_0^{(l)}, v_0^{(l)}) \in N$, $\psi_1^{(l)} \in H^1(\Omega) \cap L^\infty(\Omega)$ satisfy $\|\psi_1^{(l)}\|_{\infty,\Omega} \leqq D$, $\psi_1^{(l)} \overset{l\to\infty}{\to} \psi_1$ in $H^1(\Omega)$ and $(u_0^{(l)}, v_0^{(l)}) \overset{l\to\infty}{\to} (u_0, v_0)$ in $(L^2(\Omega))^2$. Then $(u_1^{(l)}, v_1^{(l)}) \overset{l\to\infty}{\to} (u_1, v_1)$ in $(H^1(\Omega))^2$, where $u_1^{(l)}, v_1^{(l)}$ are the solutions of (3.2.13), (3.2.14) and (3.2.15), (3.2.16) resp. with ψ_1, u_0, v_0 substituted by $\psi_1^{(l)}, u_0^{(l)}, v_0^{(l)}$.*

The continuity of the operator H is an immediate consequence of the Lemmas 3.2.4 and 3.2.5 and complete continuity follows from Lemma 3.2.3. This concludes the proof of Theorem 3.2.1. □

Similar existence proofs can be found in [3.3], [3.9], [3.11a], [3.15], [3.17], [3.19] and [3.23]. The assumptions used to prove Theorem 3.2.1 are satisfied by many commonly used models for pure semiconductor devices without Schottky-contacts. The most severe restrictions are the treatment of temperature as external quantity, the exclusion of impact ionisation in high injection situations and of velocity saturation at high electric fields.
Device models, which treat temperature as an internal quantity, were considered in [3.23], [3.24], [3.25]. The assumptions used there are very restrictive, they do not allow the modeling of 'thermal device breakdown'.
Bounds on the number of possible electron-hole carrier pairs can easily be derived from the estimates (3.2.8), (3.2.9). From (3.2.3) we obtain $np = \delta^4 uv$ and therefore

$$\delta^4 e^{-2U_+} \leqq n^*(x)p^*(x) \leqq \delta^4 e^{2U_+} \quad \text{a.e. in } \Omega \tag{3.2.22}$$

holds for $n^* = \delta^2 e^{\psi^*} u^*$, $p^* = \delta^2 e^{-\psi^*} v^*$, i.e. for the solution, whose existence was proven in Theorem 3.2.1.
We remark that the L^∞-estimates (3.2.8), (3.2.9), (3.2.10) for u, v and ψ were only proven to hold for *a* solution of the stationary semiconductor device problem. Under more stringent assumptions on the recombination-generation rate S this can be improved. For the simplest case $S \equiv 0$, i.e. under the zero-recombination-generation assumption, the maximum principle implies the estimates (3.2.8), (3.2.9) for the unique solutions $u_1 = u(\psi_1)$ and $v_1 = v(\psi_1)$ of (3.2.13), (3.2.14) and (3.2.15), (3.2.16) resp. independently of $\psi_1 \in H^1(\Omega) \cap L^\infty(\Omega)$.
Therefore (3.2.8), (3.2.9) and (3.2.10) (because of Lemma 3.2.1) hold for all solutions of the coupled problem (SD).
Also the estimates can be improved if the boundary conditions (3.2.6), which are physically reasonable for Ohmic contacts, are imposed. We set $\underline{U} := \min_{i=1,\dots,r} U_i$, $\bar{U} := \max_{i=1,\dots,r} U_i$ and assume that C is sufficiently smooth close to $\partial\Omega_D$ such that (A.3.2.2) holds. Then by choosing N as the set

$$N = \{(u,v) \in (L^2(\Omega))^2 \,|\, e^{-\underline{U}} \leqq u(x),\ v(x) \leqq e^{\bar{U}} \text{ a.e. in } \Omega\} \tag{3.2.23}$$

instead of (3.2.19) (b), the existence proof of Theorem 3.2.1 remains valid and the existence of a weak solution satisfying the L^∞-estimates

$$e^{-\bar{U}} \leqq u^*(x) \leqq e^{-\underline{U}} \quad \text{a.e. in } \Omega \tag{3.2.24}$$

$$e^{\underline{U}} \leqq v^*(x) \leqq e^{\bar{U}} \quad \text{a.e. in } \Omega \tag{3.2.25}$$

$$\ln\left[\frac{\underline{C}+\sqrt{\underline{C}^2+4\delta^4}}{2\delta^2}\right] + \underline{U} \leqq \psi^*(x)$$

$$\leqq \ln\left[\frac{\bar{C}+\sqrt{\bar{C}^2+4\delta^4}}{2\delta^2}\right] + \bar{U} \quad \text{a.e. in } \Omega \tag{3.2.26}$$

follows.
We set $|U_{\max}| = \bar{U} - \underline{U}$, i.e. $|U_{\max}|$ is the maximal absolute value of the voltages applied between any two contacts. Then (3.2.22) becomes

$$\delta^4 e^{-|U_{\max}|} \leqq n^*(x)p^*(x) \leqq \delta^4 e^{|U_{\max}|} \quad \text{a.e. in } \Omega \tag{3.2.27}$$

and estimates for the carrier concentrations follow from (3.2.3):

$$\frac{\underline{C}+\sqrt{\underline{C}^2+4\delta^4}}{2} e^{-|U_{\max}|} \leqq n^*(x)$$

$$\leqq \frac{\bar{C}+\sqrt{\bar{C}^2+4\delta^4}}{2} e^{|U_{\max}|} \quad \text{a.e. in } \Omega \tag{3.2.28}$$

$$\frac{-\bar{C}+\sqrt{\bar{C}^2+4\delta^4}}{2} e^{-|U_{\max}|} \leqq p^*(x)$$

$$\leq \frac{-\underline{C}+\sqrt{\underline{C}^2+4\delta^4}}{2} e^{|U_{\max}|} \quad \text{a.e. in } \Omega. \tag{3.2.29}$$

If $S \equiv 0$ then (3.2.24)–(3.2.29) hold for all solutions of the problem (SD). We remark that the assumption (A.3.2.5) (ii) on the mobilities can be modified to include velocity saturation for one-dimensional models. If, instead, we assume

$$0 < \underline{\mu}_n(\psi') \leq \mu_n(x, \psi') \leq \bar{\mu}_n,$$
$$0 < \underline{\mu}_p(\psi') \leq \mu_p(x, \psi') \leq \bar{\mu}_p, \qquad x \in \Omega, \qquad \psi' \in \mathbb{R}$$

with functions $\underline{\mu}_n, \underline{\mu}_p \in C(\mathbb{R})$; $\bar{\mu}_n, \bar{\mu}_p \in \mathbb{R}$, then Theorem 3.2.1 carries over since $\|\psi'\|_{\infty,\Omega}$ can be estimated in terms of U_+, $\bar{C}$ and $\underline{C}$. This implies that μ_n, μ_p are bounded away from zero uniformly for all ψ' for which the potential ψ satisfies (3.2.11), (3.2.12) with $(u_0, v_0) \in N$.
The existence of solutions (in weighted Sobolev spaces) of certain multidimensional velocity-saturation device models was shown in [3.11a].

3.3 Global Regularity

In Section 3.2 we proved that – under certain assumptions – a solution of the static semiconductor device problem, which is bounded and has square integrable first weak derivatives, exists.
When the boundary data and the parameter functions are sufficiently smooth, it is fairly easy to employ standard regularity results for (scalar) elliptic boundary value problems (see [3.10], [3.14]) in order to show that every weak solution is a classical solution, i.e. that it has continuous second derivatives in Ω, continuous first derivatives in $\Omega \cup \partial\Omega_N$ and that it assumes the Dirichlet boundary values continuously.
Singularities of solutions of multidimensional device problems can only occur at critical boundary points, that are points in $\partial\Omega$, at which Dirichlet and Neumann boundary segments meet and points, at which $\partial\Omega$ does not have a local $C^{2,\alpha}$-parametrisation.
We now turn to the investigation of global regularity of weak solutions of (SD). In particular we shall give conditions which guarantee $H^2(\Omega)$-regularity.
Even in only two space dimensions we cannot generally expect the weak solutions of (SD) to be in $H^2(\Omega)$ unless all angles, under which Neumann and Dirichlet boundary segments meet, are less than $\pi/2$ (see [3.12], [3.30], [3.31]).
We therefore have to impose rather strong conditions on the boundary data and on $\partial\Omega$. We assume:

(A.3.3.1) The Dirichlet boundary data ψ_D, u_D and v_D are in $H^2(\Omega)$ and satisfy

$$(\partial\psi_D/\partial\nu)|_{\partial\Omega_N} = (\partial u_D/\partial\nu)|_{\partial\Omega_N} = (\partial v_D/\partial\nu)|_{\partial\Omega_N} = 0.$$

The assumption (A.3.3.1) says that the data $\psi_D|_{\partial\Omega_D}$, $u_D|_{\partial\Omega_D}$, $v_D|_{\partial\Omega_D}$ can be extended smoothly to $\bar{\Omega}$ in such a way that their extensions satisfy the homogeneous Neumann-boundary conditions.
We also assume:

(A.3.3.2) The solution w of $\Delta w = f$ in Ω, $w|_{\partial\Omega_D} = (\partial w/\partial\nu)|_{\partial\Omega_N} = 0$ satisfies

$$\|w\|_{2,q,\Omega} \leqq K_1 \|f\|_{q,\Omega} \text{ for every } f \in L^q(\Omega) \text{ with } q = 2 \text{ and } q = 3/2.$$

An L^q-theory ($q > 1$) for elliptic equations can be found in [3.4], [3.10].
We remark that (A.3.3.2) actually represents a condition on the boundary segments $\partial\Omega_D$ and $\partial\Omega_N$. It excludes that Dirichlet and Neumann segments meet under angles larger than $\pi/2$. We prove:

Theorem 3.3.1: *Let the assumptions (A.3.2.1)–(A.3.2.5), (A.3.3.1), (A.3.3.2) hold and assume that* $\mu_n = \mu_n(x)$, $\mu_p = \mu_p(x)$; $\mu_n, \mu_p \in W^{1,\infty}(\Omega)$. *Then every weak solution* $(\psi, u, v) \in (H^1(\Omega) \cap L^\infty(\Omega))^3$ *satisfies*

$$(\psi, u, v) \in (H^2(\Omega))^3. \tag{3.3.1}$$

Proof: Let (ψ^*, u^*, v^*) be a weak solution. We set $\psi_r = \psi^* - \psi_D$, $u_r = u^* - u_D$, $v_r = v^* - v_D$. Then ψ_r solves

$$\lambda^2 \Delta\psi_r = -\varrho^*(x) - \lambda^2 \Delta\psi_D(x) \text{ in } \Omega, \quad \left.\frac{\partial\psi_r}{\partial\nu}\right|_{\partial\Omega_N} = \psi_r|_{\partial\Omega_D} = 0,$$

where $\varrho^* = -(\delta^2 e^{\psi^*} u^* - \delta^2 e^{-\psi^*} v^* - C) \in L^\infty(\Omega)$. Since $\psi_D \in H^2(\Omega)$, the assumption (A.3.3.2) implies ψ_r (and ψ^*) $\in H^2(\Omega)$. By carrying out the differentiation in the electron continuity equation we obtain the following problem for u_r:

$$\Delta u_r = \frac{1}{\mu_n} e^{-\psi^*} S^*(x) - \frac{1}{\mu_n} \operatorname{grad} \mu_n \cdot \operatorname{grad} u^* - \operatorname{grad} \psi^* \cdot \operatorname{grad} u^* - \Delta u_D$$

$$\left.\frac{\partial u_r}{\partial\nu}\right|_{\partial\Omega_N} = u_r|_{\partial\Omega_D} = 0,$$

where $S^*(x) := S(x, \psi^*(x), u^*(x), v^*(x)) \in L^\infty(\Omega)$, $\Delta u_D \in L^2(\Omega)$ and $\operatorname{grad} \mu_n \cdot \operatorname{grad} u^* \in L^2(\Omega)$.
Sobolev's imbedding Theorem (see [3.1]) implies $\operatorname{grad} \psi^* \in (L^6(\Omega))^k$ for $k = 1, 2, 3$ and Hölder's inequality gives $\operatorname{grad} \psi^* \cdot \operatorname{grad} u^* \in L^{\frac{3}{2}}(\Omega)$. From (A.3.3.2) ($q = 3/2$) we conclude u_r (and u^*) $\in W^{2,\frac{3}{2}}(\Omega)$. Another application of Sobolev's imbedding Theorem gives $\operatorname{grad} u^* \in (L^3(\Omega))^k$ and $\operatorname{grad} \psi^* \cdot \operatorname{grad} u^* \in L^2(\Omega)$ follows from Hölder's inequality. We obtain $u^* \in H^2(\Omega)$ from (A.3.3.2) ($q = 2$). $v^* \in H^2(\Omega)$ follows by proceeding analogously with the hole continuity equation. □

If (A.3.3.2) is only assumed to hold for $q = 2$, then $\psi \in H^2(\Omega)$ follows, but we cannot guarantee the square-integrability of all second derivatives of u and v without additional assumptions.

Obviously, (A.3.3.2) was selected with respect to the (physically reasonable) condition $k \leqq 3$. It has to be strengthened, if Theorem 3.3.1 shall be extended to the only mathematically interesting cases $k > 3$ and it can be slightly relaxed for $k = 2$.
From $(\psi, u, v) \in (H^2(\Omega))^3$ we immediately conclude by using Sobolev's imbedding Theorem:

$$(\psi, u, v) \in \left(C^{0,\frac{1}{2}}(\bar{\Omega})\right)^3 \quad \text{for} \quad k = 3 \tag{3.3.2}$$

$$(\psi, u, v) \in (C^{0,\beta}(\bar{\Omega}))^3 \quad \text{for every} \quad 0 < \beta < 1 \quad \text{if} \quad k = 2. \tag{3.3.3}$$

Although rectangular domains Ω are often used for simulations (mostly giving realistic outflow currents) the angles between insulating segments and contacts are in reality always equal to π. Therefore singularities in the electric field and in the current density components, which correspond to large electric fields and large current densities at contact edges in the 'physical' device, occur. This phenomenon is called current-crowding (see [3.8]) and is well known in device physics.

3.4 Uniqueness for Small Voltages

It is well known to semiconductor device physicists that certain devices admit multiple stationary states for certain biasing conditions. The maybe most famous example for nonuniqueness is the snap back phenomenon in thyristor technology (see [3.28]) shown in Fig. 3.4.1.
On the other hand physics tells us that a device is in thermal equilibrium, if all potentials, which are externally applied to semiconductor contacts, are zero, i.e. $J_n \equiv J_p \equiv 0$ holds then. The current relations (3.2.4) (a), (b) imply that u and v are constant in thermal equilibrium and the boundary conditions at the contacts give $u \equiv v \equiv 1$. Poisson's equation $(SD1)$ with $u \equiv v \equiv 1$ has to be solved to obtain the so called equilibrium potential $\psi = \psi_e$. Clearly $(\psi, u, v) \equiv (\psi_e, 1, 1)$ constitutes a solution of the device problem since the recombination-generation rate S vanishes identically in Ω at the equilibrium solution $(\psi_e, 1, 1)$.
Therefore it is physically reasonable to conjecture that a device admits a unique stationary state, if the potentials, which are applied to semiconductor contacts, are sufficiently small. We shall show that the equilibrium solution $(\psi, u, v) = (\psi_e, 1, 1)$ is isolated (i.e. the Frechet derivative of the device problem at the equilibrium solution is boundedly invertible), if avalanche phenomena are excluded. Then, by employing the implicit function Theorem (see [3.5], [3.13]) we shall prove that (locally) unique solutions of the semiconductor problem exist, if all applied voltages are sufficiently small and that these solutions depend continuously differentiably on the voltages.
In order to model the dependence of solutions on the externally applied potentials we write the Dirichlet conditions as:

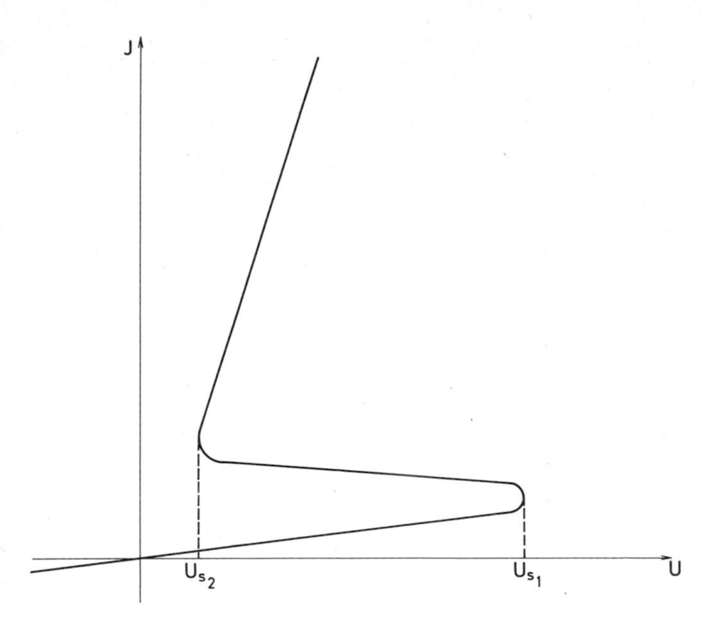

Fig. 3.4.1 Snap-back phenomenon for a thyristor

$$\psi|_{\partial\Omega_D} = \psi_D(U)|_{\partial\Omega_D},\ u|_{\partial\Omega_D} = u_D(U)|_{\partial\Omega_D},\ v|_{\partial\Omega_D} = v_D(U)|_{\partial\Omega_D}, \tag{3.4.1}$$

where $U := (U_1, \ldots, U_r)$ is the vector of the scaled potentials, which are externally applied to the r contacts.
The following assumptions will be used for the analysis:

(A.3.4.1) $\Omega \subseteq \mathbb{R}^k$ ($k = 1, 2$ or 3) is of class $C^{0,1}$, $\partial\Omega_N$ consists of C^2-segments and the $(k-1)$-dimensional Lebesgue measure of $\partial\Omega_D$ is positive.

(A.3.4.2) The mobilities are field-independent, i.e.

$$\mu_n = \mu_n(x), \quad \mu_p = \mu_p(x); \quad \mu_n, \mu_p \in W^{1,\infty}(\Omega)$$
$$0 < \underline{\mu}_n \leqq \mu_n(x) \leqq \bar{\mu}_n, \quad 0 < \underline{\mu}_p \leqq \mu_p(x) \leqq \bar{\mu}_p \quad \text{a.e. in } \Omega$$

(A.3.4.3) The recombination-generation rate S satisfies $S = F(x, \psi, u, v)\cdot(uv-1)$, where $F(x, ., ., .) \in C^2(\mathbb{R}\times(0,\infty)^2)$ for all $x \in \Omega$ and $\partial^{\alpha}_{(\psi,u,v)}F(., \psi, u, v) \in L^{\infty}(\Omega)$ uniformly for (ψ, u, v) in bounded sets of $\mathbb{R}\times(0,\infty)^2$ for every multiindex α with $|\alpha| \leqq 2$. Also $0 < \underline{\omega} \leqq F(x, \psi_e(x), 1, 1) \leqq \bar{\omega}$ or $F(x, \psi_e(x), 1, 1) \equiv 0$ in Ω holds.

(A.3.4.4) The Dirichlet data (ψ_D, u_D, v_D) form a Lipschitz-continuously differentiable map of U from $\mathbb{R}^r$ into $(H^2(\Omega))^3$. Also $(\partial\psi_D(U)/\partial\nu)|_{\partial\Omega_N} = (\partial u_D(U)/\partial\nu)|_{\partial\Omega_N} = (\partial v_D(U)/\partial\nu)|_{\partial\Omega_N} = 0$ for all $U \in \mathbb{R}^r$ and $u_D(0) = v_D(0) = 1$ in Ω.

The equilibrium problem reads:

$$\lambda^2 \Delta\psi = \delta^2(e^{\psi} - e^{-\psi}) - C(x) \quad \text{in } \Omega \tag{3.4.2}$$

$$\left.\frac{\partial \psi}{\partial \nu}\right|_{\partial\Omega_N} = 0 \tag{3.4.3}$$

$$\psi\,|_{\partial\Omega_D} = \psi_D(0)\,|_{\partial\Omega_D}. \tag{3.4.4}$$

We conclude from Lemma 3.2.1 that it is uniquely solvable. Its solution $\psi = \psi_e \in H^1(\Omega) \cap L^\infty(\Omega)$ is called equilibrium potential.
We now prove the existence and uniqueness theorem for small applied voltages.

Theorem 3.4.1: *Let the assumptions (A.3.4.1), (A.3.4.2), (A.3.4.3), (A.3.4.4), (A.3.2.3) and (A.3.3.2) hold. Then, if $|U| < \sigma$ holds for some sufficiently small σ, the problem (SD) has a locally unique solution $(\psi^*(U), u^*(U), v^*(U)) \in (H^2(\Omega))^3$, which satisfies $(\psi^*(0), u^*(0), v^*(0)) = (\psi_e, 1, 1)$ and which depends continuously differentiably on U when regarded as map from $\{U \in \mathbb{R}^r \mid |U| < \sigma\}$ into $(H^2(\Omega))^3$.*

Proof: We set $\varphi = \psi - \psi_D(U)$, $y = u - u_D(U)$, $z = v - v_D(U)$ and rewrite the problem (SD) as

$$\begin{aligned} &\lambda^2 \Delta\varphi - [\delta^2 e^{\varphi+\psi_D}(y+u_D) \\ &- \delta^2 e^{-\varphi-\psi_D}(z+v_D) - C(x) - \lambda^2 \Delta\psi_D] = 0 \end{aligned} \tag{3.4.5}$$

$$\begin{aligned} &\operatorname{div}(\mu_n e^{\varphi+\psi_D} \operatorname{grad}(y+u_D)) \\ &- F(x, \varphi+\psi_D, y+u_D, z+v_D)((y+u_D)(z+v_D)-1) = 0 \end{aligned} \tag{3.4.6}$$

$$\begin{aligned} &\operatorname{div}(\mu_p e^{-\varphi-\psi_D} \operatorname{grad}(z+v_D)) \\ &- F(x, \varphi+\psi_D, y+u_D, z+v_D)((y+u_D)(z+v_D)-1) = 0 \end{aligned} \tag{3.4.7}$$

$$\left.\frac{\partial\varphi}{\partial\nu}\right|_{\partial\Omega_N} = \left.\frac{\partial y}{\partial\nu}\right|_{\partial\Omega_N} = \left.\frac{\partial z}{\partial\nu}\right|_{\partial\Omega_N} = \varphi\,|_{\partial\Omega_D} = y\,|_{\partial\Omega_D} = z\,|_{\partial\Omega_D} = 0. \tag{3.4.8}$$

We regard the boundary value problem (3.4.5)–(3.4.8) as operator equation

$$G(\varphi, y, z, U) = 0. \tag{3.4.9}$$

The domain of G is the Cartesian product of an open subset A of $(H^2_\partial(\Omega))^3$, where $H^2_\partial(\Omega) := \{\varphi \in H^2(\Omega) \mid \partial\varphi/\partial\nu\,|_{\partial\Omega_N} = \varphi\,|_{\partial\Omega_D} = 0\}$, and a sphere $S_{\sigma_1}(0) \subseteq \mathbb{R}^r$. The set A and the radius σ_1 have to be chosen such that $u = y + u_D(U) > 0$, $v = z + v_D(U) > 0$ for all $(\varphi, y, z) \in A$ and all $U \in S_{\sigma_1}(0)$, since the function S is only defined for $u > 0$, $v > 0$ (Sobolev's imbedding Theorem implies $\|f\|_{\infty,\Omega} \leqq L\,\|f\|_{2,2,\Omega}$ for all $f \in H^2(\Omega)$, $\Omega \subseteq \mathbb{R}^k$ with $k = 1, 2$ or 3!).
By using the estimate $\|fg\|_{2,\Omega} \leqq \text{const}\,\|f\|_{1,2,\Omega} \cdot \|g\|_{1,2,\Omega}$ for $k \leqq 3$ we conclude that $G(\varphi, y, z, U) \in (L^2(\Omega))^3$ for $(\varphi, y, z) \in A$ and $U \in S_{\sigma_1}(0)$.
Thus we regard G as map:

$$G\colon\ A \times S_{\sigma_1}(0) \to (L^2(\Omega))^3, \qquad A \subseteq (H^2_\partial(\Omega))^3, \qquad S_{\sigma_1}(0) \subseteq \mathbb{R}^r \tag{3.4.10}$$

Obviously $(\psi_e - \psi_D(0), 0, 0, 0) \in A \times S_{\sigma_1}(0)$ and

$$G(\psi_e - \psi_D(0), 0, 0, 0) = 0 \tag{3.4.11}$$

holds. Also, it is easy to show that G is continuously Frechet differentiable.

In order to apply the implicit function Theorem (see [3.5], [3.13]) we show that the Frechet derivative $D_{(\varphi,y,z)}G(\psi_e-\psi_D(0),0,0,0)$ is boundedly invertible. Let $(f_1,f_2,f_3)\in(L^2(\Omega))^3$. Then the equation $D_{(\varphi,y,z)}G(\psi_e-\psi_D(0),0,0,0)\,(a,b,c)=$ $=(f_1,f_2,f_3)$ is equivalent to the boundary value problem:

$$\left.\begin{aligned}\lambda^2\Delta a &= (\delta^2 e^{\psi_e}+\delta^2 e^{-\psi_e})a+\delta^2 e^{\psi_e}b-\delta^2 e^{-\psi_e}c+f_1 \\ \operatorname{div}(\mu_n e^{\psi_e}\operatorname{grad} b) &= \omega(x)b+\omega(x)c+f_2 \\ \operatorname{div}(\mu_p e^{-\psi_e}\operatorname{grad} c) &= \omega(x)b+\omega(x)c+f_3\end{aligned}\right\}\ \text{in } \Omega \qquad \begin{aligned}&(3.4.12)\\&(3.4.13)\\&(3.4.14)\end{aligned}$$

$$\left.\frac{\partial a}{\partial \nu}\right|_{\partial\Omega_N}=\left.\frac{\partial b}{\partial \nu}\right|_{\partial\Omega_N}=\left.\frac{\partial c}{\partial \nu}\right|_{\partial\Omega_N}=a|_{\partial\Omega_D}=b|_{\partial\Omega_D}=c|_{\partial\Omega_D}=0, \tag{3.4.15}$$

where $\omega(x):=F(x,\psi_e(x),1,1)$ satisfies $0<\underline{\omega}\leqq\omega(x)\leqq\bar{\omega}$ or $\omega\equiv 0$ in Ω. The equations (3.4.13) and (3.4.14) are decoupled from (3.4.12) and they are decoupled from each other if $\omega\equiv 0$. In the latter case their unique solvability is immediate.

Now let $\omega\not\equiv 0$. We denote

$$L_1 b:=\operatorname{div}(\mu_n e^{\psi_e}\operatorname{grad} b),\qquad L_2 c=\operatorname{div}(\mu_p e^{-\psi_e}\operatorname{grad} c), \tag{3.4.16}$$

where L_1, L_2: $H_0^1(\Omega\cup\partial\Omega_N)\to(H_0^1(\Omega\cup\partial\Omega_N))^*$ (the superscript '*' denotes the dual space). Then (3.4.13), (3.4.14) subject to the boundary conditions for b and c given in (3.4.15) are equivalent to

$$\begin{bmatrix} E_1 & -(L_1-\omega E_1)^{-1}E_2\omega \\ -(L_2-\omega E_1)^{-1}E_2\omega & E_1\end{bmatrix}\begin{pmatrix} b\\ c\end{pmatrix} = \begin{pmatrix}(L_1-\omega E_1)^{-1}E_2 f_2\\ (L_2-\omega E_1)^{-1}E_2 f_3\end{pmatrix}, \tag{3.4.17}$$

where E_1 denotes the identity operator on $H_0^1(\Omega\cup\partial\Omega_N)$ and E_2 the (bounded) imbedding operator E_2: $L^2(\Omega)\to(H_0^1(\Omega\cup\partial\Omega_N))^*$. Note that $L_1-\omega E_1$ and $L_2-\omega E_1$ are boundedly invertible since $\omega(x)\geqq 0$ and since the $(k-1)$-dimensional Lebesgue measure of $\partial\Omega_D$ is positive (see [3.10], Chapter 8). The operator equation $(L_1-\omega E_1)^{-1}E_2\omega y=g$ is represented by the boundary value problem

$$\operatorname{div}(\mu_n e^{\psi_e}\operatorname{grad} g)-\omega(x)g=\omega(x)y(x),\qquad \left.\frac{\partial g}{\partial r}\right|_{\partial\Omega_N}=g|_{\partial\Omega_D}=0$$

We conclude that the estimate

$$\|g\|_{1,2,\Omega}\leqq \text{const.}\,\|y\|_{2,\Omega}$$

holds for every $y\in L^2(\Omega)$. Thus $(L_1-\omega E_1)^{-1}E_2\omega$: $L^2(\Omega)\to L^2(\Omega)$ is compact (see [3.1], Chapter 6). $(L_1-\omega E_1)^{-1}E_2$ is selfadjoint on $L^2(\Omega)$, since the differential operator L_1 is formally selfadjoint. Therefore $(L_1-\omega E_1)^{-1}E_2\omega$ is selfadjoint in the space $L^2(\Omega)$ equipped with the scalar product

$$(f,g)_\Omega^\omega:=\int_\Omega \omega(x)f(x)g(x)\,dx;\qquad f,g\in L^2(\Omega). \tag{3.4.18}$$

The induced norm $\|f\|_{2,\Omega}^\omega=\sqrt{(f,f)_\Omega^\omega}$ is equivalent to $\|f\|_{2,\Omega}$ since $\sqrt{\underline{\omega}}\,\|f\|_{2,\Omega}\leqq\|f\|_{2,\Omega}^\omega\leqq\sqrt{\bar{\omega}}\,\|f\|_{2,\Omega}$ holds.

Thus the operator $(L_1-\omega E_1)^{-1}E_2\omega$ has a bounded sequence of real eigenvalues $\chi_i \neq 0$ (see [3.6], Chapter 6) satisfying $(L_1-\omega E_1)^{-1}E_2\omega y_i = \chi_i y_i$ for some $y_i \not\equiv 0$. This implies that (μ_i, y_i) is an eigenpair of the problem

$$\operatorname{div}(\mu_n e^{\psi_e} \operatorname{grad} y_i) - \mu_i \omega(x) y_i = 0$$

$$\left.\frac{\partial y_i}{\partial \nu}\right|_{\partial\Omega_N} = y_i|_{\partial\Omega_D} = 0$$

with $\mu_i = 1+(1/\chi_i)$. The maximum principle implies $\sup \mu_i < 0$ and $-1 < \inf \chi_i$, $\chi_i < 0$ follows. Therefore we obtain the estimate

$$\|(L_1-\omega E_1)^{-1}E_2\omega\|^{\omega}_{L^2(\Omega)\to L^2(\Omega)} \leq \sup|\chi_i| < 1$$

(the superscript 'ω' means that $L^2(\Omega)$ is equipped with the norm $\|.\|^{\omega}_{2,\Omega}$). Analogously we derive:

$$\|(L_2-\omega E_1)^{-1}E_2\omega\|^{\omega}_{L^2(\Omega)\to L^2(\Omega)} < 1.$$

Banach's Lemma (see [3.10]) implies that the equation (3.4.17) is uniquely solvable and

$$\|b\|_{1,2,\Omega}+\|c\|_{1,2,\Omega} \leq \text{const.}\,(\|f_2\|_{2,\Omega}+\|f_3\|_{2,\Omega}) \tag{3.4.19}$$

follows.
By carrying out the differentiation in (3.4.13), (3.4.14), using the estimate (3.4.19) and a sequence of estimates similar to those used in the proof of Theorem 3.3.1 we obtain

$$\|b\|_{2,2,\Omega}+\|c\|_{2,2,\Omega} \leq \text{const.}\,(\|f_2\|_{2,\Omega}+\|f_3\|_{2,\Omega})$$

and

$$\|a\|_{2,2,\Omega} \leq \text{const.}\,(\|f_1\|_{2,\Omega}+\|f_2\|_{2,\Omega}+\|f_3\|_{2,\Omega})$$

follows by inserting b and c into (3.4.12).
Therefore

$$\|(D_{(\varphi,y,z)}G(\psi_e-\psi_D(0),0,0,0))^{-1}\|_{(L^2(\Omega))^3\to(H^2(\Omega))^3} \leq \text{const.}$$

holds and the implicit function theorem implies the assertion of Theorem 3.4.1. □

Because of (A.3.4.3) the theorem holds for the Shockley–Read–Hall and the Auger recombination-generation terms (and of course, for the sum of both terms). Impact ionisation is excluded. Note that we did not prove global uniqueness of the solution in $(H^2(\Omega))^3$ for small $|U|$ but only local uniqueness. For the zero-recombination-generation case $R \equiv 0$ this result can be extended, i.e. global uniqueness in $(H^2(\Omega))^3$ can be shown for sufficiently small applied biases. No global uniqueness theorem for even small applied biases is available to the author's knowledge for recombination-generation terms as general as given by (A.3.4.3). Physically, however, uniqueness of dynamically stable solutions for small biases is expected for Shockley–Read–Hall and for the Auger-type recombination-generation terms.
As expected by physical reasoning there is no uniqueness proof, which holds

for arbitrary applied potentials U. Neither is there a nonuniqueness proof for the device problem under the simplistic assumptions of Theorem 3.4.1. However, there is a numerical evidence (see [3.18]) that for one-dimensional devices, determined by doping profiles, which have at least three *pn*-junctions, global uniqueness does not even hold if $R \equiv 0$ and if the mobilities are constant..

Theorem 3.4.1 does not give bounds for the radius σ of the bias-sphere, within which local uniqueness holds. Generally, σ may depend on the domain Ω, on the doping profile C, the mobilities μ_n, μ_p and on the parameters δ and λ.

For the one-dimensional problem with two Ohmic contacts and $R \equiv 0$ Mock [3.19] showed that $\sigma \geqq \ln 2$.

Clearly, there is a variety of possible sources of nonuniqueness of stationary states such as multiple device junctions, strongly field-dependent mobilities, current-dependent temperature modeling and recombination-generation modeling, particularly avalanche generation. Numerical computations confirmed that the avalanche generation rate leads to multiple stationary states for certain reverse bias ranges even in the case of a one-dimensional *pn*-junction diode with constant mobilities and nonuniqueness for the applied bias $U = 0$ was proven for the case of (unphysically) large ionisation rates (see [3.16]).

Mock [3.19] used monotone operator theory (see [3.33]) to prove uniqueness close to thermal equilibrium.

Gajewski [3.9a] proved the uniqueness of steady state solutions under the assumptions that – for fixed boundary data ψ_D, u_D, v_D – the normed Debye length λ, whose square appears as multiplier of the Laplacian of ψ in Poisson's equation, is sufficiently large ($\lambda \gg 1$). For relevant devices, however, λ is very small ($\lambda \ll 1$) (see Chapter 4) and, thus, this result, although mathematically interesting, is of no great practical importance.

3.5 Global Continuous Branches of Solutions

In Section 3.4 we showed – assuming sufficient regularity – that locally unique solutions of the semiconductor problem exist, if all voltages applied to semiconductor contacts are sufficiently small and that these solutions depend smoothly on the applied potentials. The proof is based on showing that the equilibrium solution is isolated and therefore only gives a 'small voltage result'.

Using compactness arguments we shall now prove that the solutions of the semiconductor device equations depend in a somewhat more general sense continuously on the applied potentials assuming that the recombination-generation rate vanishes identically.

This result will be used to investigate the structure of voltage-current characteristics.

At the end of this section we shall discuss the generalisation to physically more reasonable recombination-generation rates.

The Zero-Recombination-Generation Case

We assume that the boundary data ψ_D, u_D and v_D only depend on a scalar parameter V and that the device is in thermal equilibrium for $V = 0$. Physically, this corresponds to applying zero-potentials ($U_j = 0$) to $r-1$ semiconductor contacts and varying the potential V which is applied to the r-th contact:

(A.3.5.1) The Dirichlet data $\psi_D = \psi_D(V)$, $u_D = u_D(V)$, $v_D = v_D(V)$ are continuous maps (as functions of V) from $\mathbb{R}$ into $(H^1(\Omega))^3$ and into $(L^\infty(\partial\Omega_D))^3$, $u_D(0) = v_D(0) = 1$ in Ω, $\inf_{\partial\Omega_D} u_D(V) > 0$, $\inf_{\partial\Omega_D} v_D(V) > 0$ in Ω for all $V \in \mathbb{R}$.

We denote by $(SD0)$ the problem (SD) of Section 3.2 with $R \equiv S \equiv 0$ and by $S_S \subseteq (H^1(\Omega) \cap L^\infty(\Omega))^3 \times \mathbb{R}$ the set of solutions of the problem $(SD0)$ for all voltages $V \in \mathbb{R}$, that is the set of all $(\psi, u, v, V) \in (H^1(\Omega) \cap L^\infty(\Omega))^3 \times \mathbb{R}$ for which (ψ, u, v) (with $u > 0$, $v > 0$ almost everywhere in Ω) is a solution of $(SD0)$ with boundary data $\psi_D = \psi_D(V)$, $u_D = u_D(V)$, $v_D = v_D(V)$. Also we denote by S_S^+ (S_S^-) those subsets of S_S for which the voltage parameter V is nonnegative (nonpositive).
By a continuous branch of solutions B, which meets a solution (ψ_1, u_1, v_1, V_1), we mean a subcontinuum of S_S, i.e. a closed and connected subset of S_S, where S_S is equipped with a suitable topology, containing (ψ_1, u_1, v_1, V_1).
We now state and prove the main result of this section.

Theorem 3.5.1: *Let the assumptions (A.3.2.1), (A.3.2.3), (A.3.2.4) and (A.3.4.1) hold. Then the solution subsets* $S_S^+ \subseteq (H^1(\Omega) \cap L^\infty(\Omega))^3 \times [0, \infty)$ *and* $S_S^- \subseteq (H^1(\Omega) \cap L^\infty(\Omega))^3 \times (-\infty, 0]$ *contain unbounded continuous branches* B^+ *and* B^- *resp. The branches* B^+ *and* B^- *meet the equilibrium solution* $(\psi_e, 1, 1, 0)$ *and the projections of* B^+ *and* B^- *onto the V-axis equal* $[0, \infty)$ *and* $(-\infty, 0]$ *resp., i.e.* B^+ *and* B^- *contain at least one solution for every nonnegative and nonpositive voltage resp.*

For the proof we reformulate $(SD0)$ as fixed point problem. Therefore we need the quantities

$$\bar{u}(V) := \sup_{\partial\Omega_D} u_D(V), \qquad \underline{u}(V) := \inf_{\partial\Omega_D} u_D(V).$$

$$\bar{v}(V) := \sup_{\partial\Omega_D} v_D(V), \qquad \underline{v}(V) := \inf_{\partial\Omega_D} v_D(V)$$

$$\bar{\psi}(V) := \max\left(\sup_{\partial\Omega_D} \psi_D(V),\ \ln\left[\frac{\bar{C} + \sqrt{\bar{C}^2 + 4\delta^2 \underline{u}(V)\bar{v}(V)}}{2\delta^2 \underline{u}(V)}\right]\right)$$

$$\underline{\psi}(V) := \min\left(\inf_{\partial\Omega_D} \psi_D(V),\ \ln\left[\frac{\underline{C} + \sqrt{\underline{C}^2 + 4\delta^4 \bar{u}(V)\underline{v}(V)}}{2\delta^2 \bar{u}(V)}\right]\right)$$

and

$$K(V) := \max\left(|\bar{\psi}(V)|,\ |\underline{\psi}(V)|\right). \tag{3.5.1}$$

(A.3.5.1) implies $K \in C(\mathbb{R})$.
For $\varphi \in L^2(\Omega)$, $V \in \mathbb{R}$ we set

$$\varphi_{K(V)}(x) := \begin{cases} K(V) & \text{if} \quad \varphi(x) > K(V) \\ \varphi(x) & \text{if} \quad |\varphi(x)| \leqq K(V). \\ -K(V) & \text{if} \quad \varphi(x) < -K(V) \end{cases} \tag{3.5.2}$$

The map $(\varphi, V) \to \varphi_{K(V)}$, $L^2(\Omega) \times \mathbb{R} \to L^2(\Omega)$ is continuous.
For $\varphi \in L^2(\Omega)$, $V \in \mathbb{R}$ we denote by $T_u(\varphi, V) =: u(\varphi_{K(V)}, V)$ the solution of

$$\operatorname{div}(\mu_n e^{\varphi_{K(V)}} \operatorname{grad} u) = 0, \quad \left.\frac{\partial u}{\partial \nu}\right|_{\partial\Omega_N} = 0, \quad u|_{\partial\Omega_D} = u_D(V)|_{\partial\Omega_D} \tag{3.5.3}$$

and by $T_v(\varphi, V) =: v(\varphi_{K(V)}, V)$ the solution of

$$\operatorname{div}(\mu_p e^{-\varphi_{K(V)}} \operatorname{grad} v) = 0, \quad \left.\frac{\partial v}{\partial \nu}\right|_{\partial\Omega_N} = 0, \quad v|_{\partial\Omega_D} = v_D(V)|_{\partial\Omega_D}. \tag{3.5.4}$$

By $\sigma = Z(\varphi, V)$ we denote the solution of

$$\lambda^2 \Delta\sigma = \delta^2 e^{\sigma} u(\varphi_{K(V)}, V) - \delta^2 e^{-\sigma} v(\varphi_{K(V)}, V) - C(x), \quad x \in \Omega \tag{3.5.5}$$

$$\left.\frac{\partial \sigma}{\partial \nu}\right|_{\partial\Omega_N} = 0, \quad \sigma|_{\partial\Omega_D} = \psi_D(V)|_{\partial\Omega_D}. \tag{3.5.6}$$

The maximum principle implies:

$$\underline{u}(V) \leqq u(\varphi_{K(V)}, V) \leqq \bar{u}(V), \quad \underline{v}(V) \leqq v(\varphi_{K(V)}, V) \leqq \bar{v}(V) \tag{3.5.7}$$

and, since $0 < \underline{u}(V)$, $0 < \underline{v}(V)$, we conclude from Lemma 3.2.1 that the problem (3.5.5), (3.5.6) is uniquely soluble in $H^1(\Omega) \cap L^\infty(\Omega)$ and that

$$\underline{\psi}(V) \leqq \sigma \leqq \bar{\psi}(V) \tag{3.5.8}$$

holds. Thus Z: $L^2(\Omega) \times \mathbb{R} \to L^2(\Omega)$ is well-defined and, obviously, every fixed point of $Z(., V)$ is a solution of ($SD0$) and vice versa.
Clearly $u(\varphi, 0) \equiv v(\varphi, 0) \equiv 1$ in Ω and since ψ_e is the unique solution of (3.4.2)–(3.4.4) we conclude $Z(\varphi, 0) \equiv \psi_e$.
The proof of the following Lemma is left to the reader.

Lemma 3.5.1: *Let the assumptions of Theorem 3.5.1 hold. Then the operators* Z, T_u, T_v: $L^2(\Omega) \times \mathbb{R} \to L^2(\Omega)$ *are completely continuous and* Z, T_u, T_v: $L^2(\Omega) \times \mathbb{R} \to H^1(\Omega)$ *are continuous. For every* $V \in \mathbb{R}$ *the ranges of* $Z(., V)$, $T_u(., V)$, $T_v(., V)$: $L^2(\Omega) \to L^2(\Omega)$ *are bounded.*

From Rabinowitz's fixed point Theorem (see [3.21]) we conclude the existence of unbounded continuous solution branches $B_\psi^+ \subseteq L^2(\Omega) \times [0, \infty)$, $B_\psi^- \subseteq L^2(\Omega) \times (-\infty, 0]$ of the fixed point problem $\psi = Z(\psi, V)$, which both meet the equilibrium solution $(\psi_e, 0)$.

Since $\|\psi\|_{2,\Omega} \leqq (\mu_k(\Omega))^{\frac{1}{2}} K(V)$ holds for every fixed point of $Z(., V)$, the branches B_ψ^+, B_ψ^- are unbounded 'with respect to V', i.e. they contain at least one solution for every $V \geqq 0$ and $V \leqq 0$ resp.
B_ψ^+, B_ψ^- are closed in $H^1(\Omega) \times \mathbb{R}$ because they are contained in $H^1(\Omega) \times \mathbb{R}$ and closed in $L^2(\Omega) \times \mathbb{R}$.

The continuity of Z: $L^2(\Omega) \times \mathbb{R} \to H^1(\Omega)$ implies that the identity map i: $S_\psi \subseteq L^2(\Omega) \times \mathbb{R} \to H^1(\Omega) \times \mathbb{R}$, where S_ψ denotes the solution set of the fixed-point problem $\psi = Z(\psi, V)$, is continuous. Therefore B_ψ^+, B_ψ^- are connected in $H^1(\Omega) \times \mathbb{R}$. The continuity of the maps T_u, T_v: $S_\psi \to H^1(\Omega) \times \mathbb{R}$ implies the assertion of Theorem 3.5.1. □

If the recombination-generation rate vanishes identically, then there are continuous branches of solutions B^+, B^- which emanate from the equilibrium solution and which contain solutions for all nonnegative and nonpositive voltages resp., (see Fig. 3.5.1). Note that no assertion about uniqueness is made and that bifurcation of branches is not excluded by the theorem. A restrictive assumption of the theorem in the case of a device with more than two contacts is that the potentials applied to all contacts but one are zero, i.e. $u_D(0) \equiv v_D(0) \equiv 1$ in Ω. If, instead, we assume that the potentials applied to $r-1$ semiconductor contacts are kept fixed but not necessarily zero and that the potential V, which is applied to the remaining r-th contact, varies (which corresponds to choosing $u_D(0)$, $v_D(0)$ arbitrarily), then the assertion of Theorem 3.5.1 has to be modified. It can be shown by using a generalisation of the global fixed point theorem of [3.21] (see [3.22]) that there exist unbounded continuous branches of solutions B_1^+, B_1^-, which emanate from some solution $(\psi_0, u_0, v_0, 0)$ of the problem $(SD0)$ with $V = 0$. B_1^+, B_1^- contain at least one solution for every nonnegative and nonpositive applied potential V resp.

For the case $u_D(0) \equiv v_D(0) \equiv 1$ we proved in Theorem 3.4.1 (under slightly more stringent regularity assumptions) that a unique continuously differenti-

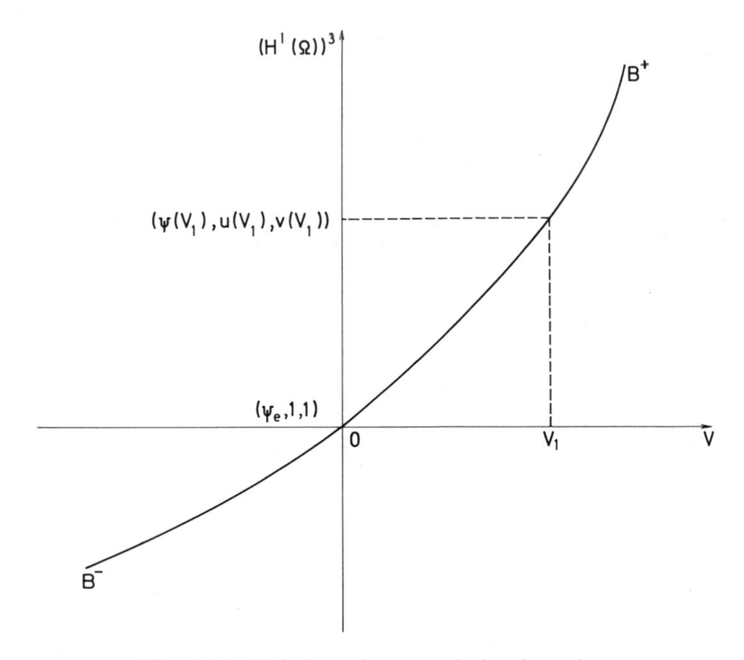

Fig. 3.5.1 Global continuous solution branches

able curve of solutions $(\psi(V), u(V), v(V), V) \in (H^2(\Omega))^3 \times \mathbb{R}$, which contains the equilibrium solution, exists for $|V|$ sufficiently small. Theorem 3.5.1 asserts that this curve can be extended unboundedly in both V-directions in such a way that the so obtained continuous solution curves $C^+ \subseteq B^+$, $C^- \subseteq B^-$ contain solutions for every $V \geqq 0$ and $V \leqq 0$ resp.

Voltage–Current Characteristics

For the device engineer the most important information about the performance of a device is given by the relationship of input voltages and outflow currents represented by so called voltage–current (V–J) characteristics. The (scaled) outflow current J_l at a semiconductor contact C_l is mathematically defined by:

$$\begin{aligned} J_l &= \int_{C_l} J \cdot \nu \, ds = \int_{C_l} (J_n + J_p) \cdot \nu \, ds \\ &= \int_{C_l} (\delta^2 \mu_n e^{\psi} \operatorname{grad} u - \delta^2 \mu_p e^{-\psi} \operatorname{grad} v) \cdot \nu \, ds. \end{aligned} \tag{3.5.9}$$

Assume now that the device under consideration has r semiconductor contacts $C_1, \ldots, C_r$ (which are closed, disjoint and – for $k = 3$ – each contact C_l has a $C^{0,1}$-boundary curve) and that all applied potentials U_j, $j = 1, \ldots, r$, except $U_i = V$ are kept fixed. Then the outflow current J_l at the contact C_l, which may be different from C_i, can be regarded as a possibly multivalued function of the input voltage V at the contact C_i. This 'function' represents the U_i–J_l-characteristic. Clearly the U_i–J_l-characteristic depends on the fixed applied potentials U_j for $j \neq i$.
In mathematical terms we define the U_i–J_l-characteristic VC_{il} for fixed U_j, $j \neq i$ as

$$\begin{aligned} VC_{il} = \{(V, J) \in \mathbb{R}^2 \mid & \text{ there is a weak solution } (\psi, u, v) \\ & \text{of the semiconductor device problem for fixed } U_j, j \neq i \\ & \text{and } U_i = V \text{ for which } J_l = J\} \subseteq \mathbb{R}^2. \end{aligned} \tag{3.5.10}$$

For analytical purposes the expression for J_l given in (3.5.9) is rather inconvenient, since the right hand side is not well defined for all $(\psi, u, v) \in (H^1(\Omega) \cap L^\infty(\Omega))^3$. We will now devise a remedy for this.
Assume that (A.3.4.1) holds. Then the distance $r_l(x)$ of a point $x \in \Omega$ to the contact C_l is a $W^{1,\infty}$-function in a sufficiently small neighbourhood of C_l. We choose a function a: $[0, \infty) \to [0, 1]$, $a \in C^\infty([0, \infty))$ which satisfies

$$a(t) = \begin{cases} 1, & 0 \leqq t < \dfrac{\omega}{2} \\ 0, & t > \omega \end{cases}$$

for $\omega > 0$ and set $\gamma_l(x) := a(r_l(x))$. Clearly $\gamma_l \in W^{1,\infty}(\Omega)$ and

$$\gamma_l|_{C_m} = \begin{cases} 1 & \text{if} \quad l = m \\ 0 & \text{if} \quad l \neq m \end{cases} \tag{3.5.11}$$

for ω small enough.
We subtract the continuity equation ($SD3$) from ($SD2$), multiply by $\delta^2\gamma_l$ and use Green's first identity:

$$\int_\Omega \gamma_l \operatorname{div} J \, dx = -\int_\Omega J \cdot \operatorname{grad} \gamma_l \, dx + \int_{\partial\Omega} \gamma_l J \cdot v \, ds = 0.$$

We obtain from (3.5.11) and from the zero-Neumann boundary conditions (3.1.7):

$$\begin{aligned} J_l &= \int_\Omega J \cdot \operatorname{grad} \gamma_l \, dx \\ &= \int_\Omega (\delta^2 \mu_n e^\psi \operatorname{grad} u - \delta^2 \mu_p e^{-\psi} \operatorname{grad} v) \cdot \operatorname{grad} \gamma_l \, dx. \end{aligned} \tag{3.5.12}$$

The outflow current J_l in the form (3.5.12) is well defined for all $\psi \in L^\infty(\Omega)$, $(u, v) \in (H^1(\Omega))^2$. If all quantities involved are sufficiently smooth, then (3.5.12) is equivalent to (3.5.9). To express the dependence of J_l on ψ, u, v we write $J_l = J_l(\psi, u, v)$.
Physically, one expects the voltage–current characteristics to consist of smooth curves in $\mathbb{R}^2$. For the zero-recombination-generation problem ($SD0$) we now prove (by using Theorem 3.5.1) that VC_{il} contains a continuous unbounded curve. We assume that the applied potentials U_j are zero for $j \neq i$ and that $U_i = V$ varies in $\mathbb{R}$.

Theorem 3.5.2: *Let the assumptions of Theorem 3.5.1 hold for the problem (SD0). Then the voltage–current-characteristic VC_{il} contains a continuous curve $\Gamma_{il} \subseteq \mathbb{R}^2$, which passes through the origin* (0, 0) *and whose projection onto the V-axis equals* $\mathbb{R}$.

Proof: Theorem 3.5.1 implies that there is a continuous curve $(\psi, u, v, V) \in (H^1(\Omega) \cap L^\infty(\Omega))^3 \times \mathbb{R}$ of solutions of ($SD0$), which passes through the equilibrium solution $(\psi_e, 1, 1, 0)$ and which is defined for all $V \in \mathbb{R}$. Let Γ_{il} be the image of this curve under the map $(\psi, u, v, V) \to (J_l(\psi, u, v), V)$, $S_S \to \mathbb{R}^2$. Since $J_l(\psi, 1, 1) = 0$, Γ_{il} passes through (0, 0). The continuity of the functional $J_l = J_l(\psi, u, v)$ as given by (3.5.12) implies the assertion. □

Note that Theorem 3.4.1 implies (under slightly more stringent regularity assumptions) that Γ_{il} is the graph of a single-valued function for sufficiently small $|V|$.
If the applied potentials U_j, $j \neq i$ are not all equal to zero, then – for the problem ($SD0$) – the existence of a continuous curve $\Gamma_{il} \subseteq VC_{il}$, which is defined for all $V \in \mathbb{R}$, is assured, but Γ_{il} does not necessarily pass through (0, 0).

Non-vanishing Recombination-Generation Terms – Impact Ionisation

The generalisation of the global results of the Theorems 3.5.1 and 3.5.2 to the semiconductor device problem with a non-vanishing recombination-generation term is nontrivial and only few results are available so far.
Even if the recombination-generation rate $S = F(uv-1)$ satisfies (A.3.4.3), we cannot generally guarantee that the continuity equations are uniquely solvable for given potential because of their strong coupling. To assure the unique solvability of the continuity equations for arbitrary given potentials and boundary data Mock [3.19] assumed that the partial derivatives $\partial S/\partial u$ and $\partial S/\partial v$ are nonnegative, bounded and continuous as functions of $u > 0$ and $v > 0$. This holds for the Shockley–Read–Hall but not for the Auger term.
Physically the performance of a device 'away from thermal equilibrium' is strongly influenced by various recombination-generation effects. Thus the practical importance of the Theorems 3.5.1 and 3.5.2 is rather limited. However, we expect the results to carry over to the semiconductor device problem (SD) with a recombination-generation term of the form (A.3.4.3) (recall the global existence Theorem 3.2.1, the local uniqueness Theorem 3.4.1 and the above mentioned result of Mock, which allows the inclusion of the Shockley–Read–Hall term).
The situation is even more complicated when impact ionisation, which is represented by the so called avalanche term:

$$R_A = -\alpha_n(\operatorname{grad}\psi)\,|J_n| - \alpha_p(\operatorname{grad}\psi)\,|J_p|, \tag{3.5.13}$$

is taken into account. Here $\alpha_n > 0$, $\alpha_p > 0$ are the (scaled) electron and hole ionisation rates resp.
Physics tells us that the impact ionisation process dominates the performance of a device when the applied potentials get large (see [3.28]). It is expected that the 'physical' continuous curve of solutions cannot be extended beyond a certain threshold voltage and that the current blows up as the applied voltage approaches the threshold voltage.
Mathematically, there is no existence proof –even for small voltages– for the multi-dimensional avalanche problem yet.
However, for a very special device, namely the one-dimensional, symmetric, and piecewise homogeneously doped diode with two Ohmic contacts in reverse bias, there is a global existence Theorem (see [3.16]).
For this device it was shown (under the restrictive assumption $\alpha_n = \alpha_p = \alpha$, where α is a nonnegative, nondecreasing and bounded function of $\|\psi'\|_{\infty,\Omega}$) that there exists an unbounded continuous branch of solutions (of the problem (SD) with the avalanche-generation rate given by (3.5.13)), which emanates from the equilibrium solution and which contains at least one solution for every negative voltage V. Thus, for this simple model problem, there is no negative threshold voltage (contrary to physical theory).
The behaviour of the voltage–current-characteristic was shown to depend heavily on the ionisation rate for large electric fields, i.e. on

$\lim_{\|\psi'\|_{\infty,\Omega}\to\infty} \alpha(\|\psi'\|_{\infty,\Omega}) =: \gamma$. For sufficiently large γ the current density J decreases exponentially to $-\infty$ as the voltage V tends to $-\infty$, while J only decreases linearly to $-\infty$ for small γ. Therefore, at least for this simple device the 'avalanche case' (γ large) is only distinguished from the 'non-avalanche case' ($\gamma = 0$ or γ small) by a more rapid increase of the absolute value of the current density as the reverse bias gets large (see Fig. 3.5.2).

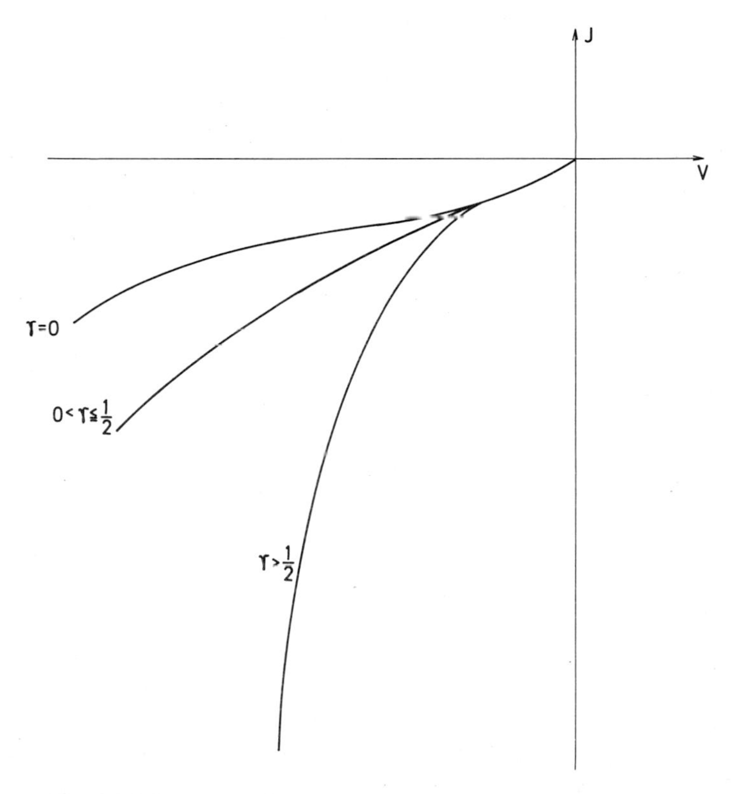

Fig. 3.5.2 Structure of the V-J-characteristic in dependence of γ

We remark that the proofs of these results 'are strictly one-dimensional', since they heavily use the explicit and unique solvability of the continuity equations (which are nonlinearly coupled because of the generation term (3.5.13)) in terms of a given potential ψ and a given current density J, and an explicit relation between the contact voltage, the potential and the current density. It is therefore not clear whether the results obtained for this simple, one-dimensional device can be extended to more complicated, multi-dimensional problems. However, the global existence result for the simple model problem suggests that avalanche generation does not solely cause junction-breakdown. It seem likely that thermal effects, which are not included in the model problem analyzed in [3.16], in conjunction with impact ionisation are responsible.

3.6 Approximate Solutions

In this section we address two questions. Firstly, we prove that functions, which approximately solve the semiconductor device problem, approximate solutions, and secondly we derive iteration schemes, which converge to the solutions of the device problem and which involve the solution of a simpler problem in every iteration step. An analysis of the same type can be found in [3.19].

The results obtained here will be heavily employed for the analysis of discretisation methods in Chapter 5.

For simplicity we mainly consider the zero recombination-generation problem ($SD0$). The dependence of the Dirichlet data ψ_D, u_D, v_D on the voltage vector U is unimportant for the subsequent analysis and will be ignored hereafter. The mobilities are assumed to be field-independent.

Basic for the following is the reformulation of the problem ($SD0$) as fixed-point equation for a compact operator as in the proof of Theorem 3.5.1. The existence Theorem 3.2.1 implies that the a-priori estimates (3.2.8)–(3.2.10) hold for every weak solution of ($SD0$). We choose $K > \max(|\bar{\psi}|, |\underline{\psi}|)$ where $\bar{\psi}$ and $\underline{\psi}$ are the upper and lower bounds resp. of ψ as given in (3.2.10) and define φ_K as in (3.5.2) and T_u, T_v, Z as functions of φ only as in (3.5.3), (3.5.4) and (3.5.5) resp. We assume:

(A.3.6.1) $\partial\Omega_N$ consists of C^2-segments, $\psi_D \in H^2(\Omega)$ and $(\partial\psi_D/\partial\nu)|_{\partial\Omega_N} = 0$. The solution w of $\Delta w = f$, $(\partial w/\partial\nu)|_{\partial\Omega_N} = w|_{\partial\Omega_D} = 0$ satisfies $\|w\|_{1,q,\Omega} \leqq E\|f\|_{2,\Omega}$ for all $f \in L^2(\Omega)$ and some $q > k$.

Sobolev's imbedding Theorem and (A.3.6.1) imply $\psi \in C^{0,1-q/k}(\bar{\Omega})$ for every fixed point ψ of Z.

We prove:

Lemma 3.6.1: *Let the assumptions of Lemma 3.5.1 and (A.3.6.1) hold. Then the operator Z: $C(\bar{\Omega}) \to C(\bar{\Omega})$ is completely continuous and the operators T_u, T_v: $C(\bar{\Omega}) \to H^1(\Omega)$ are Lipschitz-continuous. The range of Z in $C(\bar{\Omega})$ and the ranges of T_u, T_v in $H^1(\Omega)$ are bounded.*

Proof: Let $\psi_1, \psi_2 \in C(\bar{\Omega})$. Then $u(\psi_1) - u(\psi_2)$ satisfies

$$\operatorname{div}(\mu_n(e^{\psi_1} \operatorname{grad}(u(\psi_1) - u(\psi_2)))) = -\operatorname{div}(\mu_n(e^{\psi_1} - e^{\psi_2}) \operatorname{grad} u(\psi_2))$$

$$\left.\frac{(u(\psi_1) - u(\psi_2))}{\partial\nu}\right|_{\partial\Omega_N} = (u(\psi_1) - u(\psi_2))|_{\partial\Omega_D} = 0$$

and we obtain the estimate:

$$\begin{aligned}\|u(\psi_1) - u(\psi_2)\|_{1,2,\Omega} &\leqq D_1\|(e^{\psi_1} - e^{\psi_2})\,|\operatorname{grad} u(\psi_2)|\,\|_{2,\Omega}\\ &\leqq D_2\|u(\psi_2)\|_{1,2,\Omega}\,\|\psi_1 - \psi_2\|^{0,\Omega}\\ &\leqq D_3\,\|\psi_1 - \psi_2\|^{0,\Omega},\end{aligned}$$

where D_1, D_2, D_3 only depend on μ_n, Ω, $\|\psi_1\|_{\infty,\Omega}$, $\|\psi_2\|_{\infty,\Omega}$ and $\|u_D\|_{1,2,\Omega}$. Thus T_u (and also T_v) is Lipschitz continuous. We conclude from Lemma

3.2.1 that $\|Z(\varphi)\|^{0,\Omega} \leqq \max(|\underline{\psi}|, |\bar{\psi}|)$ (where $\underline{\psi}$, $\bar{\psi}$ are independent of φ) and (A.3.6.1) yields $\|Z(\varphi)\|^{0,1-q/k,\bar{\Omega}} \leqq$ const. The Rellich–Kondrachov imbedding theorem (see [3.1]) implies that the range of Z is precompact in $C(\bar{\Omega})$. The continuity of Z follows immediately from the continuity of Z as a map from $C(\bar{\Omega})$ into $H^1(\Omega)$ by using (A.3.6.1). □

Approximation Analysis by Decoupling

An immediate consequence of the complete continuity of Z: $L^2(\Omega) \to L^2(\Omega)$ (see Lemma 3.5.1) is the following

Lemma 3.6.2: *Let the assumptions of Lemma 3.5.1 hold and assume that the functions $\psi^{(n)} \in H^1(\Omega)$ are uniformly bounded in $L^\infty(\Omega)$ for $n \in \mathbb{N}$ and satisfy*

$$\psi^{(n)} - Z(\psi^{(n)}) \to 0 \quad \text{in} \quad H^1(\Omega) \quad \text{as} \quad n \to \infty. \tag{3.6.1}$$

Then there is a subsequence $\psi^{(n_l)}$ and a solution (ψ, u, v) of $(SD0)$ such that

$$(\psi^{(n_l)}, u(\psi^{(n_l)}), v(\psi^{(n_l)})) \to (\psi, u, v) \text{ in } (H^1(\Omega))^3 \text{ as } l \to \infty. \tag{3.6.2}$$

Proof: The complete continuity of Z: $L^2(\Omega) \to L^2(\Omega)$ implies the existence of a subsequence $\psi^{(n_l)} \to \psi$ in $L^2(\Omega)$ as $l \to \infty$. Then, since $\psi = Z(\psi)$, we have

$$\|\psi^{(n_l)} - \psi\|_{1,2,\Omega} \leqq \|\psi^{(n_l)} - Z(\psi^{(n_l)})\|_{1,2,\Omega} + \|Z(\psi^{(n_l)}) - Z(\psi)\|_{1,2,\Omega}$$

and the continuity of Z, T_u, T_v: $L^2(\Omega) \to H^1(\Omega)$ implies the assertion. □

If (ψ, u, v) is the unique solution of $(SD0)$, then $(\psi^{(n)}, u(\psi^{(n)}), v(\psi^{(n)}))$ converges to (ψ, u, v) in $(H^1(\Omega))^3$.
This Lemma is not very useful for practical applications, since, for given $\varphi^{(n)}$, $u(\varphi^{(n)})$ and $v(\varphi^{(n)})$ are generally not available. When, e.g., a discretisation method is employed, only approximations $u^{(n)}$, $v^{(n)}$ to $u(\varphi^{(n)})$, $v(\varphi^{(n)})$are known. To accomodate this situation, we denote by $\psi_1^{(n)}$ the solution of Poisson's equation with u and v substituted by the known functions $u^{(n)}$ and $v^{(n)}$ resp.:

$$\lambda^2 \Delta \psi_1^{(n)} = \delta^2 e^{\psi_1^{(n)}} u^{(n)} - \delta^2 e^{-\psi_1^{(n)}} v^{(n)} - C(x), \quad x \in \Omega \tag{3.6.3}$$

$$\left.\frac{\partial \psi_1^{(n)}}{\partial \nu}\right|_{\partial\Omega_N} = 0, \quad \psi_1^{(n)}\big|_{\partial\Omega_D} = \psi_D\big|_{\partial\Omega_D} \tag{3.6.4}$$

and prove (see also [3.19]):

Theorem 3.6.1: *Let the assumptions of Lemma 3.5.1 hold and assume that the triples $(\psi^{(n)}, u^{(n)}, v^{(n)}) \in H^1(\Omega) \times (L^\infty(\Omega))^2$ are uniformly bounded in $(L^\infty(\Omega))^3$ and $u^{(n)}(x) \geqq e^{-U_+}$, $v^{(n)}(x) \geqq e^{-U_+}$ a.e. in Ω for all $n \in \mathbb{N}$ and some $U_+ \in \mathbb{R}$. If $\|\psi^{(n)} - \psi_1^{(n)}\|_{1,2,\Omega} + \|u^{(n)} - u(\psi^{(n)})\|_{2,\Omega} + \|v^{(n)} - v(\psi^{(n)})\|_{2,\Omega} \to 0$ as $n \to \infty$, then there is subsequence $(\psi^{(n_l)}, u^{(n_l)}, v^{(n_l)})$ and a solution (ψ, u, v) of $(SD0)$ such that*

$$(\psi^{(n_l)}, u^{(n_l)}, v^{(n_l)}) \to (\psi, u, v) \quad \text{in} \quad H^1(\Omega) \times (L^2(\Omega))^2. \tag{3.6.5}$$

Proof: We estimate

$$\|\psi^{(n)}-Z(\psi^{(n)})\|_{1,2,\Omega} \leqq \|\psi^{(n)}-\psi_1^{(n)}\|_{1,2,\Omega}+\|\psi_1^{(n)}-Z(\psi^{(n)})\|_{1,2,\Omega}.$$

Denoting $\sigma^{(n)} = Z(\psi^{(n)})$, we obtain by subtraction:

$$\lambda^2 \Delta(\psi_1^{(n)}-\sigma^{(n)}) = \delta^2(e^{\xi^{(n)}}u(\psi^{(n)})+e^{-\eta^{(n)}}v(\psi^{(n)}))\,(\psi_1^{(n)}-\sigma^{(n)})$$
$$+e^{\psi_1^{(n)}}(u^{(n)}-u(\psi^{(n)}))-e^{-\psi_1^{(n)}}(v^{(n)}-v(\psi^{(n)})), \qquad x\in\Omega$$

$$\left.\frac{\partial(\psi_1^{(n)}-\sigma^{(n)})}{\partial\nu}\right|_{\partial\Omega_N} = (\psi^{(n)}-\sigma^{(n)})|_{\partial\Omega_D} = 0,$$

where $\xi^{(n)}(x)$ and $\eta^{(n)}(x)$ are between $\psi_1^{(n)}(x)$ and $\sigma^{(n)}(x)$. The assumptions of the theorem and a trivial extension of Lemma 3.2.1 imply $\|\psi_1^{(n)}-\sigma^{(n)}\|_{1,2,\Omega}\to 0$ as $n\to\infty$ and (3.6.5) follows by using Lemma 3.5.1. □

An L^∞-convergence statement can be obtained when (A.3.6.1) holds. Then Lemma 3.6.1 implies that a bounded sequence $\psi^{(n)}\in C(\bar\Omega)$, which satisfies $\psi^{(n)}-Z(\psi^{(n)})\to 0$ in $C(\bar\Omega)$, has a subsequence which converges in $C(\bar\Omega)$ to a fixed point ψ of Z. The L^∞-analogon of Theorem 3.6.1 reads:

Theorem 3.6.2: *Let the assumptions of Lemma 3.6.1 hold and assume that the triples* $(\psi^{(n)}, u^{(n)}, v^{(n)})\in(C(\bar\Omega))^3$ *are uniformly bounded in* $(L^\infty(\Omega))^3$ *and* $u^{(n)}(x)\geqq e^{-U_+}$, $v^{(n)}(x)\geqq e^{-U_+}$ *in* Ω *for all* $n\in\mathbb{N}$ *with some* $U_+\in\mathbb{R}$. *If* $\|\psi^{(n)}-\psi_1^{(n)}\|^{0,\Omega}+\|u^{(n)}-u(\psi^{(n)})\|_{2,\Omega}+\|v^{(n)}-v(\psi^{(n)})\|_{2,\Omega}\to 0$ *as* $n\to\infty$, *then there is a subsequence* $(\psi^{(n_l)}, u^{(n_l)}, v^{(n_l)})$ *and a solution* (ψ, u, v) *of (SD0) such that*

$$(\psi^{(n_l)}, u^{(n_l)}, v^{(n_l)})\to(\psi, u, v) \text{ in } C(\bar\Omega)\times(L^2(\Omega))^2 \text{ as } l\to\infty. \tag{3.6.6}$$

Clearly, the sequence $(\psi^{(n)}, u^{(n)}, v^{(n)})$ converges to (ψ, u, v), if the solution (ψ, u, v) of the problem $(SD0)$ is unique.
The Theorems 3.6.2 and 3.6.3 imply – roughly speaking – that a subsequence of given approximations $(\psi^{(n)}, u^{(n)}, v^{(n)})$ converges to a solution of the semiconductor device problem, if

(i) the approximations $u^{(n)}$, $v^{(n)}$ are close to solutions $u(\psi^{(n)})$, $v(\psi^{(n)})$ of the continuity equations with given potential $\psi^{(n)}$
and
(ii) the approximation $\psi^{(n)}$ of the potential is close to the solution $\psi_1^{(n)}$ of Poisson's equation with given $u^{(n)}$, $v^{(n)}$.

Thus, the convergence analysis of the semiconductor device equations with vanishing recombination-generation rate is reduced to the convergence analysis of the decoupled problem.
Estimates of the approximation error can be obtained if bounded invertibility of the Frechet derivative of I–Z at an approximate potential $\tilde\psi$ is assumed. Let $(\tilde\psi, \tilde u, \tilde v)$ be a given approximation to the solution (ψ, u, v) of $(SD0)$. We denote by $\tilde\psi_1$ the solution of the perturbed Poisson's equation

$$\lambda^2\Delta\tilde\psi_1 = \delta^2 e^{\tilde\psi_1}\tilde u-\delta^2 e^{-\tilde\psi_1}\tilde v-C(x), \qquad x\in\Omega \tag{3.6.7}$$

$$\left.\frac{\partial \tilde{\psi}_1}{\partial \nu}\right|_{\partial\Omega_N} = 0, \quad \tilde{\psi}_1|_{\partial\Omega_D} = \psi_D|_{\partial\Omega_D}. \tag{3.6.8}$$

Theorem 3.6.3: *Let the assumptions of Lemma 3.6.2 hold, let $(\tilde{\psi}, \tilde{u}, \tilde{v}) \in C(\bar{\Omega}) \times \times (L^\infty(\Omega))^2$ satisfy $\tilde{u}(x) \geqq e^{-U_+}$, $\tilde{v}(x) \geqq e^{-U_+}$ almost everywhere in Ω for some $U_+ \in \mathbb{R}$ and assume that the Frechet derivative $I - D_\psi Z(\tilde{\psi})$: $C(\bar{\Omega}) \to C(\bar{\Omega})$ of $I - Z$ at $\tilde{\psi}$ is boundedly invertible. Then there are constants $M_1, M_2, \chi > 0$ and an isolated solution (ψ^*, u^*, v^*) of (SD0) such that*

$$\|\psi^* - \tilde{\psi}\|^{0,\Omega} \leqq 2M_1(\|\tilde{\psi} - \tilde{\psi}_1\|^{0,\Omega} + M_2\delta^2(\|e^{\tilde{\psi}_1}\|^{0,\Omega}\,\|u(\tilde{\psi}) - \tilde{u}\|_{2,\Omega}$$
$$+ \|e^{-\tilde{\psi}_1}\|^{0,\Omega}\,\|v(\tilde{\psi}) - \tilde{v}\|_{2,\Omega})) \tag{3.6.9}$$

holds, if the right hand side of (3.6.9) is less than χ.

A solution $(\psi^*, u^*, v^*) = (\psi^*, u(\psi^*), v(\psi^*))$ is called isolated if the Frechet derivative $I - D_\psi Z(\psi^*)$ is boundedly invertible.

Proof: We define the operator $Q := I - Z$: $C(\bar{\Omega}) \to C(\bar{\Omega})$ and compute

$$Q(\tilde{\psi}) = \tilde{\psi} - Z(\tilde{\psi}) = (\tilde{\psi} - \tilde{\psi}_1) - (Z(\tilde{\psi}) - \tilde{\psi}_1).$$

Clearly $Z(\tilde{\psi}) = \tilde{\sigma}$ solves the problem

$$\lambda^2 \Delta\tilde{\sigma} = \delta^2 e^{\tilde{\sigma}} u(\tilde{\psi}) - \delta^2 e^{-\tilde{\sigma}} v(\tilde{\psi}) - C(x)$$
$$\left.\frac{\partial \tilde{\sigma}}{\partial \nu}\right|_{\partial\Omega_N} = 0, \quad \tilde{\sigma}|_{\partial\Omega_D} = \psi_D|_{\partial\Omega_D}.$$

Subtracting (3.6.7), (3.6.8) gives:

$$\lambda^2 \Delta(\tilde{\sigma} - \tilde{\psi}_1) = \delta^2(e^{\xi} u(\tilde{\psi}) + e^{-\eta} v(\tilde{\psi}))\,(\tilde{\sigma} - \tilde{\psi}_1) + f(x), \qquad x \in \Omega$$
$$\left.\frac{\partial(\tilde{\sigma} - \tilde{\psi}_1)}{\partial \nu}\right|_{\partial\Omega_N} = 0, \quad (\tilde{\sigma} - \tilde{\psi}_1)|_{\partial\Omega_D} = 0$$

with $f = \delta^2 e^{\tilde{\psi}_1}(u(\tilde{\psi}) - \tilde{u}) - \delta^2 e^{-\tilde{\psi}_1}(v(\tilde{\psi}) - \tilde{v})$, where $\xi(x)$ and $\eta(x)$ are between $\tilde{\psi}_1(x)$ and $\tilde{\sigma}(x)$. (A.3.6.1) implies $\|\tilde{\sigma} - \tilde{\psi}_1\|^{0,\Omega} \leqq M_2\|f\|_{2,\Omega}$ and the implicit function theorem applied to Q implies the assertion. □

For the analysis of discretisation schemes in the singular perturbation framework of Chapter 4 it is important to have more explicit information on the constants M_1, M_2 and χ. By going through the proof of the implicit function Theorem we find that we can choose

$$M_1 = \|(I - D_\psi Z(\tilde{\psi}))^{-1}\|_{C(\bar{\Omega}) \to C(\bar{\Omega})} \tag{3.6.10}$$
$$\chi = 1/((M_3 + 1)\,(2M_1 + 1)), \tag{3.6.11}$$

where M_3 is a local Lipschitz constant of $D_\psi Z$ in a neighbourhood of $\tilde{\psi}$. M_2 is an upper bound for the norm of the operator W: $L^2(\Omega) \to C(\bar{\Omega})$, where $Wf = w$ is defined by

$$\lambda^2 \Delta w = \delta^2(e^{\xi} u(\tilde{\psi}) + e^{-\eta} v(\tilde{\psi}))w + f, \quad \left.\frac{\partial w}{\partial \nu}\right|_{\partial\Omega_N} = w|_{\partial\Omega_D} = 0 \tag{3.6.12}$$

with $\xi(x)$ and $\eta(x)$ between $\tilde{\psi}_1(x)$ and $\tilde{\sigma}(x)$.

For the error in approximating u and v we obtain

$$\|u(\psi^*)-\tilde{u}\|_X \leqq \|\tilde{u}-u(\tilde{\psi})\|_X + M_u\|\tilde{\psi}-\psi^*\|^{0,\Omega} \tag{3.6.13}$$

$$\|v(\psi^*)-\tilde{v}\|_X \leqq \|\tilde{v}-v(\tilde{\psi})\|_X + M_v\|\tilde{\psi}-\psi^*\|^{0,\Omega}, \tag{3.6.14}$$

where X is some Banachspace of functions for which the maps T_u, T_v: $C(\bar{\Omega}) \to X$ are Lipschitz continuous with Lipschitz constants M_u and M_v resp. (e.g. $X = H^1(\Omega)$, $X = L^2(\Omega)$ or $X = C^1(\bar{\Omega})$ for one space dimension).
In most practical cases the bounded invertibility of $I - D_\psi Z(\tilde{\psi})$ is difficult to verify. Even without this condition we conclude from Theorem 3.6.2 that there exists a solution (ψ^*, u^*, v^*) of $(SD0)$ which is close to $(\tilde{\psi}, \tilde{u}, \tilde{v})$, if the right hand side of (3.6.9) is sufficiently small. If, instead of assuming that the linearisation of $I-Z$ at $\tilde{\psi}$ is boundedly invertible, we require ψ^* to be an isolated solution of $\psi - Z(\psi) = 0$, then the estimate (3.6.9) holds with

$$M_1 \approx M_1^* := \|(I - D_\psi Z(\psi^*))^{-1}\|_{C(\bar{\Omega})\to C(\bar{\Omega})}, \tag{3.6.15}$$

$$\chi \approx \chi^* := 1/((M_3^*+1)\,(2M_1^*+1)), \tag{3.6.16}$$

where M_3^* is a local Lipschitz-constant of $D_\psi Z$ at ψ^*.
Sloppily speaking, we conclude that the error estimate (3.6.9) holds at isolated solutions and the constants involved depend only on the local conditioning of the problem.

Iteration Schemes

When the semiconductor device equations are discretised, one is left with the problem of numerically solving the nonlinear system of equations produced by the discretisation scheme. To do so, one has to use some kind of iteration method, that is an algorithm which, starting from an initial guess, generates a sequence of approximations recursively, which converges to a solution of the already discretised problem.
In this subsection we shall discuss iterative methods'on the continuous level', that means we shall investigate iteration algorithms applied to the continuous semiconductor device problem. These algorithms can in an analogous way be applied to discretisation schemes or, conversely, the 'continuous' iteration scheme can be discretised appropriately (see, e.g. [3.2]).
The maybe most prominent iteration scheme is Newton's method (see [3.29]). For a Frechet-differentiable operator $F{:}X \to Y$ it can be written in the form

$$D_x F(x_l)\,(x_{l+1} - x_l) = -F(x_l),\; l = 0, 1, 2, \ldots \tag{3.6.17}$$

with given initial guess $x_0 \in X$. If the solution x^* of $F(x) = 0$ is isolated, and if F is Lipschitz continuously Frechet differentiable, then Newton's method converges quadratically to x^* for all x_0 which are sufficiently close to x^*, i.e. the estimate

$$\|x_{l+1} - x_l\|_X \leqq \text{const.}\, \|x_l - x_{l-1}\|_X^2, \qquad \lim_{l\to\infty} x_l = x^* \tag{3.6.18}$$

holds, whenever $\|x_0 - x^*\|_X < \chi$ for some χ sufficiently small (see [3.19a]). Note that the increment $\Delta x_l = x_{l+1} - x_l$ is obtained by solving the (linear) operator equation (3.6.17).

We write the problem *(SD)* of Section 3.2 in operator form $F(\psi, u, v) = 0$ (where the boundary conditions are incorporated in F) and assume for simplicity that the mobilities are field-independent and that the (smooth) recombination-generation rate only depends on ψ, u, v and x, i.e. $S = S(x, \psi, u, v)$.

Then the $(l+1)$-st step of Newton's method (3.6.17) is represented by the linear elliptic boundary value problem

$$\lambda^2 \Delta \psi_{l+1} - (\delta^2 e^{\psi_l} u_l + \delta^2 e^{-\psi_l} v_l)(\psi_{l+1} - \psi_l)$$
$$= \delta^2 e^{\psi_l} u_{l+1} - \delta^2 e^{-\psi_l} v_{l+1} - C(x) \tag{3.6.19}$$

$$\operatorname{div}(\mu_n e^{\psi_l}(\operatorname{grad} u_{l+1} + (\psi_{l+1} - \psi_l)\operatorname{grad} u_l))$$
$$= \frac{\partial S}{\partial \psi}(x, \psi_l, u_l, v_l)(\psi_{l+1} - \psi_l) + \frac{\partial S}{\partial u}(x, \psi_l, u_l, v_l)(u_{l+1} - u_l)$$
$$+ \frac{\partial S}{\partial v}(x, \psi_l, u_l, v_l)(v_{l+1} - v_l) + S(x, \psi_l, u_l, v_l) \tag{3.6.20}$$

$$\operatorname{div}(\mu_p e^{-\psi_l}(\operatorname{grad} v_{l+1} - (\psi_{l+1} - \psi_l)\operatorname{grad} v_l))$$
$$= \frac{\partial S}{\partial \psi}(x, \psi_l, u_l, v_l)(\psi_{l+1} - \psi_l) + \frac{\partial S}{\partial u}(x, \psi_l, u_l, v_l)(u_{l+1} - u_l)$$
$$+ \frac{\partial S}{\partial v}(x, \psi_l, u_l, v_l)(v_{l+1} - v_l) + S(x, \psi_l, u_l, v_l) \tag{3.6.21}$$

$$\left.\frac{\partial \psi_{l+1}}{\partial \nu}\right|_{\partial\Omega_N} = \left.\frac{\partial u_{l+1}}{\partial \nu}\right|_{\partial\Omega_N} = \left.\frac{\partial v_{l+1}}{\partial \nu}\right|_{\partial\Omega_N} = 0 \tag{3.6.22}$$

$$\psi_{l+1}|_{\partial\Omega_D} = \psi_D|_{\partial\Omega_D},\ u_{l+1}|_{\partial\Omega_D} = u_D|_{\partial\Omega_D},\ v_{l+1}|_{\partial\Omega_D} = v_D|_{\partial\Omega_D} \tag{3.6.23}$$

for $l = 0, 1, 2, \ldots$ where the initial guess (ψ_0, u_0, v_0) is given. We obtain

Theorem 3.6.4: *Let the assumptions (A.3.4.1), (A.3.4.2), (A.3.2.3), (A.3.3.1), (A.3.3.2) hold, assume that* $S(x, ., ., .) \in C^2(\mathbb{R}x(0, \infty)^2)$ *for all* $x \in \Omega$ *and* $\partial^\alpha S(x, ., ., .) \in L^\infty(\Omega)$ *uniformly in bounded subsets of* $\mathbb{R}x(0, \infty)^2$ *for all multiindices* α *with* $|\alpha| \leqq 2$. *If* $(\psi^*, u^*, v^*) \in (H^2(\Omega))^3$ *is an isolated solution of the problem (SD), then Newton's method (3.6.19)–(3.6.23) is quadratically convergent in* $(H^2(\Omega))^3$ *to* (ψ^*, u^*, v^*) *for all starting values* (ψ_0, u_0, v_0), *which are sufficiently close (in* $(H^2(\Omega))^3$*) to* (ψ^*, u^*, v^*).

If the recombination-generation term S satisfies (A.3.4.3), then Theorem 3.4.1 implies that there is a neighbourhood of the equilibrium solution $(\psi_e, 1, 1)$ within which *(SD)* has an isolated solution assuming that (ψ_D, u_D, v_D) is close to $(\psi_e, 1, 1)$ in $(H^2(\Omega))^3$. Thus, Newtons's method converges (locally) quadratically to an isolated solution of the semiconductor device problem for biasing conditions close to thermal equilibrium.

The semiconductor device problem in the (ψ, u, v)-variables is in most cases not suitable for numerical computations because of eventually occuring over/under-flow in the evaluation of the exponentials. Since the (ψ, n, p)-variables are mostly used for simulations, we also formulate Newton's method in these variables, i.e. the operator F is now represented by the problem (3.1.1)–(3.1.5) subject to mixed Neumann-Dirichlet boundary conditions. We obtain

$$\lambda^2 \Delta \psi_{l+1} = n_{l+1} - p_{l+1} - C(x) \tag{3.6.24}$$

$$\operatorname{div}\left(\mu_n(\operatorname{grad} n_{l+1} - n_{l+1} \operatorname{grad} \psi_l - n_l \operatorname{grad}(\psi_{l+1} - \psi_l))\right)$$
$$= \frac{\partial R}{\partial \psi}(x, \psi_l, n_l, p_l)(\psi_{l+1} - \psi_l) + \frac{\partial R}{\partial n}(x, \psi_l, n_l, p_l)(n_{l+1} - n_l)$$
$$+ \frac{\partial R}{\partial p}(x, \psi_l, n_l, p_l)(p_{l+1} - p_l) + R(x, \psi_l, n_l, p_l) \tag{3.6.25}$$

$$\operatorname{div}\left(\mu_p(\operatorname{grad} p_{l+1} + p_{l+1} \operatorname{grad} \psi_l + p_l \operatorname{grad}(\psi_{l+1} - \psi_l))\right)$$
$$= \frac{\partial R}{\partial \psi}(x, \psi_l, n_l, p_l)(\psi_{l+1} - \psi_l) + \frac{\partial R}{\partial n}(x, \psi_l, n_l, p_l)(n_{l+1} - n_l)$$
$$+ \frac{\partial R}{\partial p}(x, \psi_l, n_l, p_l)(p_{l+1} - p_l) + R(x, \psi_l, n_l, p_l) \tag{3.6.26}$$

$$\left.\frac{\partial \psi_{l+1}}{\partial \nu}\right|_{\partial\Omega_N} = \left.\frac{\partial n_{l+1}}{\partial \nu}\right|_{\partial\Omega_N} = \left.\frac{\partial p_{l+1}}{\partial \nu}\right|_{\partial\Omega_N} = 0 \tag{3.6.27}$$

$$\psi_{l+1}|_{\partial\Omega_D} = \psi_D|_{\partial\Omega_D},\ n_{l+1}|_{\partial\Omega_D} = n_D|_{\partial\Omega_D},\ p_{l+1}|_{\partial\Omega_D} = p_D|_{\partial\Omega_D} \tag{3.6.28}$$

for $l = 0, 1, 2, \ldots$ where ψ_0, n_0, p_0 are given.
The main difficulty in numerically solving (3.6.24)–(3.6.28) lies in the coupling of the equations. Various block iteration algorithms have been proposed as a remedy (see [3.27]).
Most often so called Gummel-type blockiteration methods (suggested by H. K. Gummel in [3.11]) are used. A typical member of this class of methods is obtained by forcibly decoupling (3.6.19), (3.6.20), (3.6.21) and by dropping all terms in the continuity equations, which contain $\psi_{l+1} - \psi_l$, $u_{l+1} - u_l$, $v_{l+1} - v_l$. Given (ψ_l, u_l, v_l) we solve:

$$\operatorname{div}\left(\mu_n e^{\psi_l} \operatorname{grad} u_{l+1}\right) = S(x, \psi_l, u_l, v_l) \tag{3.6.29}$$

$$\left.\frac{\partial u_{l+1}}{\partial \nu}\right|_{\partial\Omega_N} = 0, \qquad u_{l+1}|_{\partial\Omega_D} = u_D|_{\partial\Omega_D} \tag{3.6.30}$$

for u_{l+1} and

$$\operatorname{div}\left(\mu_p e^{-\psi_l} \operatorname{grad} v_{l+1}\right) = S(x, \psi_l, u_l, v_l) \tag{3.6.31}$$

$$\left.\frac{\partial v_{l+1}}{\partial \nu}\right|_{\partial\Omega_N} = 0, \qquad v_{l+1}|_{\partial\Omega_D} = v_D|_{\partial\Omega_D} \tag{3.6.32}$$

for v_{l+1}. The so obtained functions u_{l+1}, v_{l+1} are used to solve (3.6.19) subject to the boundary conditions $(\partial\psi_{l+1}/\partial v)|_{\partial\Omega_N} = 0$, $\psi_{l+1}|_{\partial\Omega_D} = \psi_D|_{\partial\Omega_D}$. If $S = F(x, \psi, u, v)(uv - 1)$, then $F(x, \psi_l, u_l, v_l)(u_{l+1}v_l - 1)$ can be taken as right hand side of (3.6.29) and $F(x, \psi_l, u_{l+1}, v_l)(u_{l+1}v_{l+1} - 1)$ as right hand side of (3.6.31) without destroying the linearity of the problems (3.6.29), (3.6.30) and (3.6.31), (3.6.32). For small recombination-generation and for small grad u and grad v this iteration scheme can be interpreted as small perturbation of Newton's method, in fact, for $S \equiv 0$ and $\psi_D \equiv \psi_e$, $u_D \equiv \equiv v_D \equiv 1$, that is in thermal equilibrium, the two methods coincide.
Another Gummel-type iteration is obtained by dropping the term $(\delta^2 e^{\psi_l}u_l + \delta^2 e^{-\psi_l}v_l)\cdot(\psi_{l+1} - \psi_l)$ on the left hand side of (3.6.19) and by substituting ψ_{l+1} for ψ_l on the right hand side of (3.6.19). For given (ψ_l, u_l, v_l) we solve:

$$\operatorname{div}(\mu_n e^{\psi_l}\operatorname{grad} u_{l+1}) = F(x, \psi_l, u_l, v_l)(u_{l+1}v_l - 1) \tag{3.6.33}$$

$$\left.\frac{\partial u_{l+1}}{\partial v}\right|_{\partial\Omega_N} = 0, \qquad u_{l+1}|_{\partial\Omega_D} = u_D|_{\partial\Omega_D} \tag{3.6.34}$$

for u_{l+1} and

$$\operatorname{div}(\mu_p e^{-\psi_l}\operatorname{grad} v_{l+1}) = F(x, \psi_l, u_{l+1}, v_l)(u_{l+1}v_{l+1} - 1) \tag{3.6.35}$$

$$\left.\frac{\partial v_{l+1}}{\partial v}\right|_{\partial\Omega_N} = 0, \qquad v_{l+1}|_{\partial\Omega_D} = v_D|_{\partial\Omega_D} \tag{3.6.36}$$

for v_{l+1}. Then we solve

$$\lambda^2\Delta\psi_{l+1} = \delta^2 e^{\psi_{l+1}}u_{l+1} - \delta^2 e^{-\psi_{l+1}}v_{l+1} - C(x) \tag{3.6.37}$$

$$\left.\frac{\partial\psi_{l+1}}{\partial v}\right|_{\partial\Omega_N} = 0, \qquad \psi_{l+1}|_{\partial\Omega_D} = \psi_D|_{\partial\Omega_D} \tag{3.6.38}$$

to obtain ψ_{l+1}. The continuity equations are linear in u_{l+1} and v_{l+1} resp.; however, a semilinear Poisson's equation has to be solved in each step. For practical computations the iteration (3.6.33)–(3.6.38) is usually rewritten in terms of ψ_{l+1}, $n_{l+1} := \delta^2 e^{\psi_l}u_{l+1}$, $p_{l+1} := \delta^2 e^{-\psi_l}v_{l+1}$. We obtain:

$$\operatorname{div}(\mu_n(\operatorname{grad} n_{l+1} - n_{l+1}\operatorname{grad}\psi_l)) = R_l^{(1)} \tag{3.6.39}$$

$$\operatorname{div}(\mu_p(\operatorname{grad} p_{l+1} + p_{l+1}\operatorname{grad}\psi_l)) = R_l^{(2)} \tag{3.6.40}$$

$$\lambda^2\Delta\psi_{l+1} = e^{\psi_{l+1}-\psi_l}n_{l+1} - e^{-(\psi_{l+1}-\psi_l)}p_{l+1} - C(x), \tag{3.6.41}$$

where $R_l^{(1)}$ and $R_l^{(2)}$ are the recombination-generation terms in (3.6.33) and (3.6.35) resp. expressed in the (ψ, n, p) variables.
Newton's method is normally used to solve (3.6.41) approximately. When only one step of Newton's method starting with the initial guess ψ_l is performed, then we come up with the iteration scheme, which was originally suggested by Gummel in [3.11]:

$$\lambda^2\psi_{l+1} - (n_{l+1} + p_{l+1})(\psi_{l+1} - \psi_l) = n_{l+1} - p_{l+1} - C(x) \tag{3.6.42}$$

preceded by (3.6.39), (3.6.40).

For vanishing recombination-generation rate Mock [3.19] proved convergence of iteration schemes of the form

$$\lambda^2 \Delta\psi_{l+1} - r(x)(\psi_{l+1} - \psi_l) = n_{l+1} - p_{l+1} - C(x)$$

(supplemented by (3.6.39), (3.6.40)) for a certain class of stabilising functions r, which does generally not include (3.6.42). He used methods from monotone operator theory, which, under stringent regularity requirements, are applicable close to thermal equilibrium.
The convergence of the iteration (3.6.33)–(3.6.38) can be proven constructively for vanishing recombination-generation rate under biasing conditions sufficiently close to thermal equilibrium. The proof is based on the observation that – for $R \equiv 0$ – the iteration corresponds to iterating the fixed point operator Z of Section 3.5 (see also Lemma 3.6.1), which can be shown to be contractive close to thermal equilibrium.

Theorem 3.6.5: *Let the assumptions (A.3.2.2), (A.3.2.3), (A.3.4.1), (A.3.4.2), (A.3.6.1) hold and assume that* $R \equiv 0$. *Then there are constants* $A > 0$, $0 < L < 1$, $\chi > 0$ *such that the sequence* ψ_l *generated by the iteration (3.6.33)–(3.6.38) converges linearly with factor* L *in* $W^{1,q}(\Omega)$ *to an isolated solution* ψ^* *of* $\psi - Z(\psi) = 0$, *if the following two conditions are satisfied:*

$$\|\psi_0\|_{1,q,\Omega} \leqq A \tag{3.6.43 (a)}$$

$$\|u_D - 1\|_{1,2,\Omega} + \|v_D - 1\|_{1,2,\Omega} < \chi. \tag{3.6.43 (b)}$$

Linear convergence with factor L means

$$\begin{aligned}\|\psi_{l+1} - \psi_l\|_{1,q,\Omega} &\leqq L\|\psi_l - \psi_{l-1}\|_{1,q,\Omega} \\ &\leqq \ldots \leqq L^l \|\psi_1 - \psi_0\|_{1,q,\Omega}.\end{aligned} \tag{3.6.44}$$

Proof: Obviously $u_{l+1} = T_u(\psi_l) = u(\psi_l)$, $v_{l+1} = T_v(\psi_l) = v(\psi_l)$ holds. Thus, $\psi_{l+1} = Z(\psi_l)$ follows (the operators Z, T_u, T_v are defined in Section 3.5, the voltage-dependence is suppressed here). We shall now estimate the Frechet-derivative of Z as map from $W^{1,q}(\Omega)$ into $W^{1,q}(\Omega)$.
For $\varphi, \omega \in W^{1,q}(\Omega)$ we set $\sigma = Z(\varphi)$ and $\gamma = D_\varphi Z(\varphi)\omega$. By linearising (3.5.5), (3.5.6) we obtain the following boundary value problem for γ:

$$\begin{aligned}\lambda^2 \Delta\gamma = {} & (\delta^2 e^\sigma u(\varphi) + \delta^2 e^{-\sigma} v(\varphi))\gamma \\ & + \delta^2 e^\sigma D_\varphi u(\varphi)\omega - \delta^2 e^{-\sigma} D_\varphi v(\varphi)\omega\end{aligned} \tag{3.6.45 (a)}$$

$$\gamma|_{\partial\Omega_D} = \left.\frac{\partial\gamma}{\partial\nu}\right|_{\partial\Omega_N} = 0. \tag{3.6.45 (b)}$$

From the a-priori bounds of $u(\varphi)$, $v(\varphi)$ and σ and from (A.3.6.1) we derive the estimate:

$$\|\gamma\|_{1,q,\Omega} \leqq K_1(\|D_\varphi u(\varphi)\omega\|_{2,\Omega} + \|D_\varphi v(\varphi)\omega\|_{2,\Omega}), \tag{3.6.46}$$

where K_1 depends on Ω, λ, δ, $\|u_D\|_{\infty,\partial\Omega_D}$, $\|v_D\|_{\infty,\partial\Omega_D}$, $\|\psi_D\|_{\infty,\partial\Omega_D}$, $\|C\|_{\infty,\Omega}$ and on the constant E of (A.3.6.1).

The function $\alpha_1 := D_\varphi u(\varphi)\omega$ satisfies

$$\operatorname{div}(\delta^2\mu_n e^\varphi \operatorname{grad}\alpha_1 + J_n(\varphi)\omega) = 0, \quad \alpha_1|_{\partial\Omega_D} = \left.\frac{\partial\alpha_1}{\partial\nu}\right|_{\partial\Omega_N} = 0, \quad (3.6.47)$$

where we denoted $J_n(\varphi) := \delta^2\mu_n e^\varphi \operatorname{grad} u(\varphi)$. The estimate

$$\begin{aligned}\|\alpha_1\|_{1,2,\Omega} &\leqq D_1 \| |J_n(\varphi)| \|_{2,\Omega} \|\omega\|_{\infty,\Omega} \\ &\leqq K_2 \| |J_n(\varphi)| \|_{2,\Omega} \|\omega\|_{1,q,\Omega} \qquad (3.6.48)\end{aligned}$$

follows immediately and, setting $\alpha_2 := D_\varphi v(\varphi)\omega$, $J_p(\varphi) := -\delta^2\mu_p e^{-\varphi} \operatorname{grad} v(\varphi)$, we obtain by proceeding analogously:

$$\|\alpha_2\|_{1,2,\Omega} \leqq K_3 \| |J_p(\varphi)| \|_{2,\Omega} \|\omega\|_{1,q,\Omega} \qquad (3.6.49)$$

By evoking (3.6.46) we derive:

$$\|D_\varphi Z(\psi)\|_{W^{1,q}(\Omega) \to W^{1,q}(\Omega)} \leqq K_3(\| |J_n(\varphi)| \|_{2,\Omega} + \| |J_p(\varphi)| \|_{2,\Omega}). \qquad (3.6.50)$$

We rewrite the continuity equation as:

$$\operatorname{div}(\mu_n e^\varphi \operatorname{grad}(u(\varphi)-1)) = 0,$$

$$(u(\varphi)-1)|_{\partial\Omega_D} = (u_D - 1)|_{\partial\Omega_D}, \qquad \left.\frac{\partial(u(\varphi)-1)}{\partial\nu}\right|_{\partial\Omega_N} = 0$$

and estimate

$$\| |J_n(\varphi)| \|_{2,\Omega} \leqq K_4 \|u_D - 1\|_{1,2,\Omega}. \qquad (3.6.51)\ (a)$$

Analogously we derive

$$\| |J_p(\varphi)| \|_{2,\Omega} \leqq K_5 \|v_D - 1\|_{1,2,\Omega}, \qquad (3.6.51)\ (b)$$

and (3.6.50) implies

$$\|D_\varphi Z(\varphi)\|_{W^{1,q}(\Omega) \to W^{1,q}(\Omega)} \leqq K_6(\|u_D - 1\|_{1,2,\Omega} + \|v_D - 1\|_{1,2,\Omega}). \qquad (3.6.52)$$

where K_6 depends on λ, Ω, δ, μ_n, μ_p, $\|u_D\|_{\infty,\partial\Omega_D}$, $\|v_D\|_{\infty,\partial\Omega_D}$, $\|C\|_{\infty,\Omega}$, $\|\varphi\|_{\infty,\Omega}$ and on the constant E of (A.3.6.1).
We define the closed sphere $S_\beta := \{\varphi \in W^{1,q}(\Omega) \mid \|\varphi\|_{1,q,\Omega} \leqq \beta\}$ with $\beta > (E/\lambda^2)\mu_k(\Omega)^{\frac{1}{2}}(\delta^2 e^{\bar\psi}\bar u + \delta^2 e^{-\underline{\psi}}\bar v + \|C\|_{\infty,\Omega})$, where $\bar\psi$, $\underline{\psi}$, $\bar u$, $\bar v$ denote the a-priori bounds of σ, $u(\varphi)$ and $v(\varphi)$ resp. (see (3.5.7), (3.5.8)). Then $Z\colon S_\beta \to S_\beta$ holds and $K_6 = K_6(\beta)$ for $\varphi \in S_\beta$. Taylor's Theorem gives

$$Z(\varphi_1) - Z(\varphi_2) = D_\varphi Z(\xi)(\varphi_1 - \varphi_2), \qquad (3.6.53)$$

where ξ is on the line segment connecting φ_1 and φ_2. We choose $A = \beta$ and $0 < \chi < (1/K_6(A))$. Then Z is contractive in the sphere $S_A \subseteq W^{1,q}(\Omega)$ since (3.6.52), (3.6.43) (b) imply:

$$\|Z(\varphi_1) - Z(\varphi_2)\|_{1,q,\Omega} \leqq K_6(A)\chi \|\varphi_1 - \varphi_2\|_{1,q,\Omega}; \quad \varphi_1, \varphi_2 \in S_A. \qquad (3.6.54)$$

We set $L = K_6(A)\chi$ and conclude the assertion of Theorem 3.6.5 from the contraction mapping theorem (see [3.10]). □

The error estimate

$$\|\psi_l - \psi^*\|_{1,q,\Omega} \leqq L^l \|\psi_0 - \psi^*\|_{1,q,\Omega} \leqq 2AL^l \tag{3.6.55}$$

follows from the contraction mapping theorem, too.
Lemma 3.6.1 implies that (u_l, v_l) converges to $(u(\psi^*), v(\psi^*))$ in $(H^1(\Omega))^2$.
A convergence proof for Gummel-type iterations, which requires 'less regularity', was presented in [3.12a], 'plausibility arguments' for convergence were given in [3.29] and the convergence of Gummel-type iterations applied to the appropriately discretised one-dimensional device problem was proven in [3.29]. All these results are based on a contraction-type argument similar to the proof of the Theorem 3.6.5 and therefore only hold 'close to thermal equilibrium'.
The estimate (3.6.50) indicates that the speed of convergence, determined by the factor L, which is locally (in first approximation) given by the norm of the Frechet-derivative of Z at the solution, depends decisively on the magnitude of the current densities, i.e.

$$L \approx \text{const.} (\| |J_n| \|_{2,\Omega} + \| |J_p| \|_{2,\Omega}). \tag{3.6.56}$$

This is confirmed by computational experience. Gummel-type methods converge reasonably fast (even for rather bad initial guesses), if the current densities and the recombination-generation rate are not too large, they usually converge very slowly in high injection simulations.

References

[3.1] Adams, R. A.: Sobolev Spaces. New York–San Francisco–London: Academic Press 1975.

[3.2] Bank, R. E., Rose, D. J., Fichtner, W.: Numerical Methods for Semiconductor Device Simulation. IEEE Transactions on Electron Devices *ED–30,* No. 9, 1031–1041 (1983).

[3.3] Bank, R. E., Jerome, J. W., Rose, D. J.: Analytical and Numerical Aspects of Semiconductor Modelling. Report 82–11274–2, Bell Laboratories, 1982.

[3.4] Bers, L., John, F., Scheckter, M.: Partial Differential Equations, American Mathematical Society. New York: J. Wiley 1964.

[3.5] Chow, S. N., Hale, J. K.: Methods of Bifurcation Theory. New York–Heidelberg–Berlin: Springer 1982.

[3.6] Dunford, N., Schwartz, J. T.: Linear Operators, Part 1. New York: J. Wiley 1957.

[3.7] Fichera, G.: Linear Elliptic Differential Systems and Eigenvalue Problems. (Lecture Notes in Mathematics, Vol. 8.) Berlin–Heidelberg–New York: Springer 1965.

[3.8] Franz, G. A., Franz, A. F., Selberherr, S., Ringhofer, C., Markowich, P. A.: Current Crowding Effects in Thyristors – A Two-Dimensional Numerical Analysis. Report, Institut für Allgemeine Elektrotechnik, Technische Universität Wien, Austria, 1984.

[3.9] Gajewski, H.: On the Existence of Steady-State Carrier Distributions in Semiconductors. ZAMM, Vol. 65, pp. 101–108, 1985.

[3.9a] Gajewski, H.: On Uniqueness and Stability of Steady-State Carrier Distributions in Semiconductors. (To appear in Proc. Equadiff. Conf. 1985.) Berlin–Heidelberg–New York: Springer 1985.

[3.10] Gilbarg, D., Trudinger, N. S.: Elliptic Partial Differential Equations of Second Order, 2nd ed. Berlin–Heidelberg–New York: Springer 1984.

[3.11] Gummel, H. K.: A Self-Consistent Iterative Scheme for One-Dimensional Steady State Transistor Calculations. IEEE Transactions on Electron Devices *ED–11,* 455–465 (1964).

[3.11a] Jerome, J. W.: Consistency of Semiconductor Modeling: An Existence/Stability Analysis for the Stationary Van Roosbrook System. SIAM J. Appl. Math. *45*, 565–590 (1985).
[3.12] Kawohl, B.: Über nichtlineare gemischte Randwertprobleme für elliptische Differentialgleichungen zweiter Ordnung auf Gebieten mit Ecken. Dissertation, TH Darmstadt, BRD, 1978.
[3.12a] Kerkhoven, T.: On the Dependence of the Convergence of Gummel's Algorithm on the Regularity of the Solutions. Research Report YALEU/DCS/RR–366, Yale University, Department of Computer Science, U.S.A., 1985.
[3.13] Krasnoselskii, M. A., Zabreiko, P. P.: Geometrical Methods of Nonlinear Analysis. Berlin–Heidelberg–New York: Springer 1984.
[3.14] Ladyzhenskaja, O. A., Uraltseva, N. N.: Linear and Quasilinear Elliptic Equations. New York: Academic Press 1968.
[3.15] Markowich, P. A.: A Singular Perturbation Analysis of the Fundamental Semiconductor Device Equations. SIAM J. Appl. Math. *5*, No. 44, 896–928 (1985).
[3.16] Markowich, P. A.: A Nonlinear Eigenvalue Problem Modelling the Avalanche Effect in Semiconductor Diodes. SIAM J. Math. Anal. (1985).
[3.17] Mock, M. S.: On Equations Describing Steady State Carrier Distributions in a Semiconductor Device. Comm. Pure and Appl. Math. *25*, 781–792 (1972).
[3.18] Mock, M. S.: An Example of Nonuniqueness of Stationary Solutions in Semiconductor Device Models. COMPEL *1*, 165–174 (1982).
[3.19] Mock, M. S.: Analysis of Mathematical Models of Semiconductor Devices. Dublin: Boole Press 1983.
[3.19a] Ortega, J. M., Rheinboldt, W. C.: Iterative Solution of Nonlinear Equations in Several Variables. New York–London: Academic Press 1970.
[3.20] Protter, M. H., Weinberger, H. F.: Maximum Principles in Differential Equations. Englewood Cliffs, N. J.: Prentice-Hall 1967.
[3.21] Rabinowitz, P. H.: Some Global Results for Nonlinear Eigenvalue Problems. Functional Analysis, *7*, 487–513 (1971).
[3.22] Rabinowitz, P. H.: A Global Theorem for Nonlinear Eigenvalue Problems and Applications. In: Contributions to Nonlinear Functional Analysis (Zarantonello, E. H., ed.). New York–London: Academic Press 1971.
[3.23] Seidmann, T. I.: Steady State Solutions of Diffusion-Reaction Systems with Electrostatic Convection. Nonlinear Analysis *4*, No. 3, 623–637 (1980).
[3.24] Seidmann, T. I., Troianello, G. M.: Time Dependent Solutions of a Nonlinear System Arising in Semiconductor Theory. Nonlinear Analysis (1985).
[3.25] Seidmann, T. I.: Time Dependent Solutions of a Nonlinear System Arising in Semiconductor Theory II: Boundedness and Periodicity. (To appear, 1985.)
[3.26] Seidmann, T. I., Choo, S. C.: Iterative Schemes for Computer Simulation of Semiconductor Devices. Solid-State Electronics, *15*, 1229–1235 (1972).
[3.27] Selberherr, S.: Analysis and Simulation of Semiconductor Devices. Wien–New York: Springer 1984.
[3.28] Sze, M. S.: Physics of Semiconductor Devices. New York: J. Wiley 1981.
[3.29] Watanabe, D. S., Sheikh, Q. M., Slamet, S.: Convergence of Quasi-Newton Methods for Semiconductor Equations. Report, Department of Comp. Science, University of Illinois, Urbana, Illinois, U.S.A., 1984.
[3.30] Wigley, N. M.: Mixed Boundary Value Problems in Plane Domains with Corners. Math. Z. *115*, 33–52 (1970).
[3.31] Wigley, N. M.: Asymptotic Expansions at a Corner of Solutions of Mixed Boundary Value Problems. Mathematics and Mechanics *13*, No. 4, 549–576 (1964).
[3.32] Wloka, J.: Partielle Differentialgleichungen. Stuttgart: Teubner 1982.
[3.33] Zarantonello, E. H.: Solving Functional Equations by Contractive Averaging. MRC–TSR 160, Math. Res. Center, University of Wisconsin-Madison, U.S.A., 1960.

4 Singular Perturbation Analysis of the Stationary Semiconductor Device Problem

In this chapter we present an approach to the analysis of the qualitative and quantitative structure of solutions of the stationary basic semiconductor device equations. The approach is entirely based on singular perturbation theory.
It is well known to device physicists that the potential and carrier concentrations behave totally different in certain well defined regions of a device. The gradients of these quantities are generally large in thin strips about junctions (i.e. surfaces across which the doping profile varies rapidly), Schottky contacts and semiconductor-oxide interfaces, while they vary moderately away from these surfaces. The following questions arise immediately from this observation:

- What is the underlying mathematical mechanism generating such a solution structure?
- Can this mechanism be used to characterize the solutions of the semiconductor device problem locally in 'smooth' regions and in regions of fast variation?

Clearly much mathematical and practical insight can be gained by answering these questions. Also, as we shall show in the next chapter, the structural information on the solutions can be efficiently used to assess the applicability and performance of discretisation methods for the semiconductor device equations.

4.1 The Device Equations As Singular Perturbation Problem

The scaled stationary device equations read:

$$\lambda^2 \Delta\psi = n-p-C(x) \tag{4.1.1}$$

$$\operatorname{div} J_n = R \tag{4.1.2}$$

$$\operatorname{div} J_p = -R \tag{4.1.3}$$

$$J_n = \mu_n(\operatorname{grad} n - n \operatorname{grad} \psi) \tag{4.1.4}$$

$$J_p = -\mu_p(\operatorname{grad} p + p \operatorname{grad} \psi). \tag{4.1.5}$$

These equations hold in the scaled semiconductor domain $\Omega \subseteq \mathbb{R}^k$. We assume that the semiconductor boundary $\partial\Omega$ consists of Neumann segments, whose union we denote by $\partial\Omega_N$, contact segments $\partial\Omega_D$ and – in the case of a MOS-device – an oxide-semiconductor interface $\partial\Omega_I$. $\partial\Omega_D$ splits into r_O Ohmic contacts denoted by O_i and $r-r_O$ Schottky contacts S_j. The corresponding boundary and interface conditons are given is Section 2.3 and the scaling, which leads to the formulation (4.1.1)–(4.1.5) is discussed in Section 2.4.

The parameter λ of (4.1.1) is of paramount importance for the following analysis. We derived in Section 2.4:

$$\lambda^2 = \left(\frac{\lambda_D}{l}\right)^2 = \frac{\varepsilon_s U_T}{l^2 q \tilde{C}}. \tag{4.1.6}$$

l is a characteristic length of the device chosen such that the diameter of the domain, which the semiconductor occupies, is of the order of magnitude l. $\tilde{C}$ is a characteristic doping concentration, i.e. it is of the same order of magnitude as the maximal doping concentration. l and $\tilde{C}$ are the scaling factors of the independent variable and of the doping profile resp.

Physically, λ_D is a Debye length of the device under consideration (see [4.29]), thus λ is a normed Debye length. Note that λ decreases when $\tilde{C}$ is increased and when l is increased.

For a silicon device at temperature $T = 300\ K$ with a characteristic length $l = 5\times 10^{-3}$ cm and a maximal doping concentration $\tilde{C} = 10^{17}\ \text{cm}^{-3}$ we compute $\lambda \approx 10^{-\frac{7}{2}}$. If $\tilde{C}$ is increased to $10^{20}\ \text{cm}^{-3}$ then $\lambda \approx 10^{-5}$. In most realistic applications λ is a small parameter, i.e. $\lambda \ll 1$.

The boundary and interface conditions are independent of λ, however, the conditions imposed on the contact segments $\partial\Omega_D$ depend on another parameter. At an Ohmic contact, for example, the boundary conditions read

$$\psi|_{O_i} = \psi_{\text{bi}}|_{O_i} + U_i, \qquad (n-p-C)|_{O_i} = 0, \qquad np|_{O_i} = \delta^4, \tag{4.1.7}$$

where the built-in potential ψ_{bi} is given by

$$\psi_{\text{bi}}(x) = \ln\left[\frac{C(x)+\sqrt{C(x)^2+4\delta^4}}{2\delta^2}\right]. \tag{4.1.8}$$

U_i denotes the scaled potential, which is externally applied to O_i. In Section 2.4 we derived

$$\delta^2 = \frac{n_i}{\tilde{C}}, \tag{4.1.9}$$

where n_i is the intrinsic number of the semiconductor. For a silicon device at roomtemperature with a maximal doping $\tilde{C} = 10^{20}\ \text{cm}^{-3}$ we calculate $\delta \approx 10^{-5}$. Clearly $\delta \ll 1$ holds in most practical applications, too.

Note that we obtain:

$$\psi_{\mathrm{bi}}\big|_{O_i} \sim \begin{cases} \ln \dfrac{1}{\delta^2} & \text{if} \quad C\big|_{O_i} > 0 \\ -\ln \dfrac{1}{\delta^2} & \text{if} \quad C\big|_{O_i} < 0 \end{cases} \qquad \text{as} \quad \delta \to 0+.$$

The built-in-potential blows up logarithmically as the characteristic doping concentration $\tilde{C}$ increases. The boundary data for n and p at Ohmic contacts remain bounded as $\delta \to 0+$, actually

$$n\big|_{O_i} \to \begin{cases} C\big|_{O_i} & \text{if}\, C\big|_{O_i} > 0 \\ 0 & \text{if}\, C\big|_{O_i} < 0 \end{cases}, \quad p\big|_{O_i} \to \begin{cases} 0 & \text{if}\, C\big|_{O_i} > 0 \\ -C\big|_{O_i} & \text{if}\, C\big|_{O_i} < 0 \end{cases} \qquad \text{as } \delta \to 0+$$

holds.

When one faces a mathematical problem which depends on small parameters, one is tempted to set these parameters or a certain selection of them to zero and solve the so obtained 'reduced' problem to obtain an approximation of the solution of the 'full' parameter-dependent problem.

In the semiconductor device problem the small parameters are λ and δ. Clearly, setting $\delta = 0$ does not make sense since it corresponds to prescribing infinitely large boundary data for the potential at contacts. Moreover, since λ appears as a multiplier of the Laplacian in Poisson's equation, we expect it to carry more 'structural' information on the solutions than δ, which only determines the magnitude of the built-in-potential. Therefore we will regard δ as fixed and investigate the structure of the solutions of the semiconductor device problem for sufficiently small values of λ.

The reduced equations are thus obtained by setting $\lambda = 0$. Then (4.1.1) becomes the condition of zero-space-charge in Ω:

$$0 = n - p - C(x), \qquad x \in \Omega \tag{4.1.10}$$

and (4.1.2)–(4.1.5) remain unchanged. Since λ does not appear in the boundary conditions, it is at the first glance intriguing to prescribe them for the reduced equations, too. At Ohmic contacts the boundary data satisfy the zero-space-charge condition (4.1.10), while the boundary data at Schottky contacts generally contradict (4.1.10). By computing the gradient of the right hand side of (4.1.10) we notice that the zero-space-charge condition contradicts the homogeneous Neumann boundary conditions for ψ, n, p at those points of $\partial\Omega_N$ at which $\operatorname{grad} C \cdot v\big|_{\partial\Omega_N} \neq 0$ holds.

The reason for these contradictions is that by going from (4.1.1) to (4.1.10) we loose one degree of freedom in imposing boundary conditions. We can therefore – in general – not expect the solutions of the reduced equations to satisfy all boundary conditions prescribed for the 'full' equations.

Another difficulty originates from the structure of the physical doping profile at device junctions. Mostly the doping concentration is modeled as a function which varies rapidly across junctions. To make life easier we assume for the moment that we only have abrupt junctions in our device, i.e. $C(x)$ has jump-discontinuities everywhere at each junction. Thus, an abrupt junction is – mathematically speaking – a $(k-1)$ – dimensional smooth surface $\Gamma \subseteq \Omega$ with

$$[C(x)]_\Gamma \neq 0 \qquad \text{for each } x \in \Gamma.$$

Clearly (4.1.10) implies that every solution of the reduced problem has jump-discontinuities at the junctions, too (more accurately, at least one of the reduced solutions n or p has a jump). This stems from the loss of the regularizing property of the inverse of Laplace's operator, which does not appear in (4.1.10) since it is multiplied by λ^2 in Poisson's equation (4.1.1). Every solution of the 'full' semiconductor device problem, however, is continuous in $\dot{\Omega}$ (see Section 3.3).

These considerations clearly demonstrate (they even prove) that the solutions of the reduced problem ($\lambda = 0$) do not in general approximate a solution of the semiconductor device problem uniformly in Ω to arbitrarily small accuracy if only λ is sufficiently small.

The loss of 'one degree of freedom' in imposing boundary conditions and the loss of regularity of solutions when going to the reduced problem is typical for boundary value problems in which the small parameter multiplies the highest order derivative. This indicates that the solutions have singularities at least locally in x as the perturbation parameter tends to zero. Therefore such a boundary value problem is called singularly perturbed.

The semiconductor device problem thus constitutes a singularly perturbed elliptic system and the perturbation parameter is a normed Debye length of the device under consideration.

The obvious question to ask now is whether we have to abandon the approach of setting λ equal to zero in (4.1.1) completely or whether we can define a 'reasonable' reduced problem, whose solutions are close to the solutions of the 'full' singularly perturbed device problem at least away from the critical surfaces, i.e. away from the boundary $\partial\Omega$ and from junctions. Semiconductor device physics tells us that the space charge density is small away from steep junctions, Schottky contacts and semiconductor-oxide interfaces (assuming that the applied potentials are moderate). Thus it should be possible to exploit mathematically the zero-space-charge-condition to obtain local approximations of the solutions of the singularly perturbed semiconductor device problem.

The next question is: what do the solutions of the 'full' problem look like close to the critical surfaces? Physics tells us that the potential and the carrier concentrations are steep, fast varying functions about junctions, Schottky contacts and semiconductor-oxide interfaces.

To give a mathematical description of this behaviour we shall present the approach of matched asymptotic expansions in the small paramter λ and apply it to the semiconductor device problem. This will lead to the derivation of appropriate boundary conditions for the reduced equations and to the derivation of so called layer equations, whose solutions 'characterize' the 'full' solutions close to the critical surfaces. These layer equations are second order ordinary differential equations posed on infinite intervals whose independent variables represent the stretched fast scale on which the potential and the carrier concentrations vary close to the critical surfaces. Away from these surfaces the 'full' solutions are moderately varying functions approximated by the reduced solutions.

The singular perturbation character of the semiconductor device problem

causes the solutions to vary on two different scales locally, namely on a fast one in thin regions about the boundary and about junctions and on a slow one away from these surfaces. Therefore the singular perturbation approach mathematically explains the occurence of depletion and semiconductor-oxide interface layers. Besides giving an explanation it also allows the separate qualitative and quantitative description of the solutions inside and outside the thin strips of fast variation resp. This is done by analysing the behaviour of the solutions of the reduced problem and of the layer problems, which are of considerably simpler structure than the original singularly perturbed semiconductor device problem.
The reader who wants to get deeper into the area of singular perturbations is referred to [4.4], [4.12] for excellent introductions based on physical problems. The text [4.13] is likely to appeal to mathematically oriented readers. A completely rigorous treatment of singular perturbation theory can be found in [4.6], [4.9].
The singular perturbation approach was applied to the semiconductor device equations by many authors. The one-dimensional case is dealt with in [4.16], [4.17], [4.21], [4.23], [4.28], [4.30], [4.31], higher dimensional problems in [4.18], [4.19], [4.20] and a survey of results can be found in [4.26].

4.2 A Singularly Perturbed Second Order Model Problem

In this section we shall study a simple singularly perturbed second order ordinary differential equation in order to demonstrate the concepts, which will be used thereafter to analyse the semiconductor device equations in the framework of singular perturbation theory.
The model problem reads:

$$\lambda^2\varphi'' = a^2\varphi - C(x), \qquad -1 \leqq x \leqq 1 \tag{4.2.1}$$

$$\varphi'(-1) = \alpha, \qquad \varphi(1) = \beta \tag{4.2.2}$$

with $a > 0$, $\alpha, \beta \in \mathbb{R}$ and $\lambda \in (0, \lambda_0]$ for some $0 < \lambda_0 \ll 1$. The function $C(x)$ is assumed to be piecewise constant:

$$C(x) = \begin{cases} C_+, x \in (X, 1] \\ C_-, x \in [-1, X) \end{cases}, \qquad C_+ \neq C_- \tag{4.2.3}$$

with $X \in (-1, 1)$.
The equation (4.2.1) can be regarded as a linearized 'version' of Poisson's equation. The boundary point $x = 1$ represents a contact, $x = -1$ an 'insulating' boundary point and $C(x)$ the doping profile. The hypothetical device has an abrupt junction at $x = X$. Note that φ'' is discontinuous at $x = X$. A (weak) solution of (4.2.1) satisfies $\varphi \in C^1([-1, 1])$, i.e. the conditions

$$\varphi(X+) = \varphi(X-), \qquad \varphi'(X+) = \varphi'(X-) \tag{4.2.4}$$

hold, where we denote the one-sided limits by $\varphi(X\pm) := \lim_{x \to X\pm} \varphi(x)$.

Structure of Solutions

We solve the problem (4.2.1), (4.2.2) explicitely and obtain

$$\varphi(x) = \frac{C(x)}{a^2} + \begin{cases} \frac{1}{2a^2}(C_- - C_+)e^{-\frac{a}{\lambda}|x-X|} + \left(\beta - \frac{C_+}{a^2}\right)e^{-\frac{a}{\lambda}(1-x)}, & x \in (X, 1] \\ \frac{1}{2a^2}(C_+ - C_-)e^{-\frac{a}{\lambda}|x-X|} - \frac{\lambda\alpha}{a}e^{-\frac{a}{\lambda}(x+1)}, & x \in [-1, X) \end{cases}$$

$$+ f(x, \lambda), \tag{4.2.5}$$

where $f(x, \lambda)$ together with all its x-derivatives is exponentially small for small λ, i.e.

$$|f^{(i)}(x, \lambda)| \leqq D_i e^{-\frac{b_i}{\lambda}}, \quad x \in [-1, 1];$$
$$b_i, D_i > 0 \text{ for } i = 0, 1, \ldots \tag{4.2.6 (a)}$$

holds. Using Landau symbols we write:

$$f^{(i)}(x, \lambda) = O\left(e^{-\frac{b_i}{\lambda}}\right). \tag{4.2.6 (b)}$$

The first term on the right hand side of (4.2.5), namely $\bar{\varphi}(x) := C(x)/a^2$, is the solution of the reduced equation obtained by setting λ equal to zero in (4.2.1):

$$0 = a^2\bar{\varphi}(x) - C(x). \tag{4.2.7}$$

The second term:

$$\hat{\varphi}\left(\frac{x-X}{\lambda}\right) := \frac{1}{2a^2}(C_- - C_+)\begin{cases} e^{-\frac{a}{\lambda}|x-X|}, & x \in (X, 1] \\ -e^{-\frac{a}{\lambda}|x-X|}, & x \in [-1, X) \end{cases} \tag{4.2.8}$$

causes the solution φ to lift off from the reduced solution $\bar{\varphi}$ close to the junction $x = X$. It makes φ continuous at $x = X$ by connecting the two 'branches' of the discontinuous reduced solution $\bar{\varphi}(x)$.
The third term:

$$\tilde{\varphi}(x, \lambda) = \begin{cases} \tilde{\varphi}_+\left(\frac{1-x}{\lambda}\right) := \left(\beta - \frac{C_+}{a^2}\right)e^{-\frac{a}{\lambda}(1-x)}, & x \in (X, 1] \\ \tilde{\varphi}_-\left(\frac{x+1}{\lambda}\right) := -\frac{\lambda\alpha}{a}e^{-\frac{a}{\lambda}(x+1)}, & x \in [-1, X) \end{cases} \tag{4.2.9}$$

forces φ to stay away from $\bar{\varphi}$ close to the boundaries $x = \pm 1$, thus allowing φ to assume the prescribed boundary conditions. Figure 4.2.1 depicts the structure of a typical solution φ.
$\hat{\varphi}$ and $\tilde{\varphi}_+$ are of significant magnitude only in small intervals about the junction $x = X$ and the boundary point $x = 1$ resp. These transition intervals, within which the solution φ is not approximated to order λ by the reduced solution, are called (zeroth order) layers or (zeroth order) layer

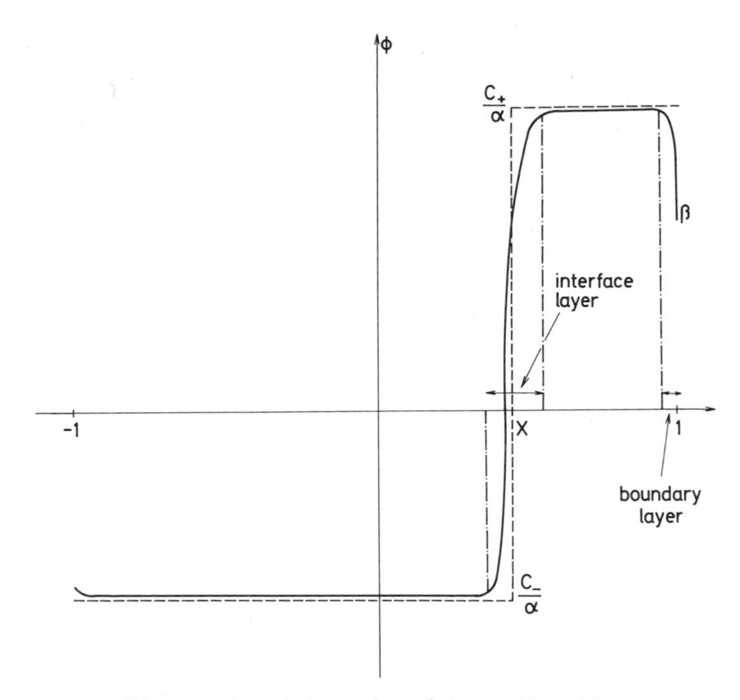

Fig. 4.2.1 A typical solution of the model problem

intervals, and the terms $\hat{\varphi}$ and $\tilde{\varphi}_+$, which cause the lift-off from the reduced solution, are called (zeroth order) junction and boundary layer terms resp. Note that $\tilde{\varphi}_-$ is maximally $O(\lambda)$.
Outside the layers the solution φ is a slowly varying function of x approximated uniformly to order λ by the reduced solution and inside each layer it is an exponentially fast varying function of x approximated by the sum of the corresponding boundary layer term and the reduced solution evaluated at the boundary point and, resp., by the sum of the junction-layer term and the respective one-sided limit of the reduced solution at the junction.
Because of this the reduced solution is also called zeroth order outer term and the layer solutions zeroth order inner terms.
A zeroth order layer is an interval, within which a zeroth order inner term is in absolute value larger than a 'moderate' multiple of the perturbation parameter λ. For the junction layer at $x = X$ we obtain with $d > 0$:

$$\left|\hat{\varphi}\left(\frac{x-X}{\lambda}\right)\right| \geqq d\lambda \leftrightarrow |x-X| \leqq \frac{\lambda}{a}\left|\ln\frac{|C_- - C_+|}{2a^2 d\lambda}\right|. \tag{4.2.10}$$

With $d \approx 1$ we compute the width $w_J(\lambda)$ of the junction layer at $x = X$:

$$w_J(\lambda) \approx \frac{2}{a}\lambda\left|\ln\frac{|C_- - C_+|}{2a^2 d\lambda}\right|. \tag{4.2.11}$$

Clearly $w_J(\lambda) = O(\lambda|\ln\lambda|)$ holds for $\lambda \to 0+$.
At the Dirichlet boundary we have

$$\left|\tilde{\varphi}_+\left(\frac{1-x}{\lambda}\right)\right| \geqq d\lambda \leftrightarrow |1-x| \leqq \frac{\lambda}{a}\left|\ln \frac{\left|\beta-\frac{C_+}{a^2}\right|}{d\lambda}\right| \tag{4.2.12}$$

and the width $w_D(\lambda)$ of the boundary layer at $x = 1$ is given by

$$w_D(\lambda) \approx \frac{\lambda}{a}\left|\ln \frac{\left|\beta-\frac{C_+}{a^2}\right|}{\lambda}\right|. \tag{4.2.13}$$

Thus $w_D(\lambda) = O(\lambda|\ln \lambda|)$ as $\lambda \to 0+$, too.
No layer occurs at $x = 1$, if the prescribed boundary value β is attained by the reduced solution, since then $\tilde{\varphi}_+ \equiv 0$ holds.
The Neumann boundary term $\tilde{\varphi}_-$ satisfies:

$$\left|\tilde{\varphi}_-\left(\frac{x+1}{\lambda}\right)\right| \leqq \lambda\frac{|\alpha|}{a}, \qquad x \in [-1, X). \tag{4.2.14}$$

A zeroth order layer does not occur at the Neumann boundary point $x = -1$ since the right hand side of (4.2.14) is $O(\lambda)$.
The layer heights can be computed easily. The 'jump' at the junction layer is given by

$$h_J \approx |\bar{\varphi}(X+)-\bar{\varphi}(X-)| = \left|\frac{C_- - C_+}{a^2}\right|, \tag{4.2.15}$$

and the jump at the Dirichlet boundary layer

$$h_D \approx |\varphi(1)-\bar{\varphi}(1)| = \left|\beta-\frac{C_+}{a^2}\right|. \tag{4.2.16}$$

Within the layers the derivatives of the solution blow up as λ tends to zero, e.g.:

$$|\varphi^{(i)}(x)| \approx \left(\frac{a}{\lambda}\right)^i\left|\beta-\frac{C_+}{a^2}\right| e^{-\frac{a}{\lambda}(1-x)} \tag{4.2.17}$$

holds inside the Dirichlet boundary layer. At the boundary point $x = 1$ we have

$$|\varphi^{(i)}(1)| \approx \left(\frac{a}{\lambda}\right)^i\left|\beta-\frac{C_+}{a^2}\right|, \tag{4.2.18}$$

that means $\varphi^{(i)}(1)$ is of the order of magnitude λ^{-i}, and at the points $x_i = 1-\frac{i}{a}\lambda|\ln \lambda|$, which are on the edge of the layer:

$$|\varphi^{(i)}(x_i)| \approx a^i\left|\beta-\frac{C_+}{a^2}\right|. \tag{4.2.19}$$

The i-th derivative of the solution decays exponentially from $O(\lambda^{-i})$ to $O(1)$ inside a layer.
Close to the Neumann boundary point $x = -1$ we obtain:

$$|\varphi^{(i)}(x)| \approx \alpha \left(\frac{a}{\lambda}\right)^{i-1} e^{-\frac{a}{\lambda}(x+1)}, \qquad i = 1, 2, \dots . \tag{4.2.20}$$

Thus, unless $\alpha = 0$, all derivatives of order larger than 1 blow up close to $x = -1$ as $\lambda \to 0+$.
The interval, within which the first derivative of the reduced solution does not approximate the first derivative of the solution to order λ, while the reduced solution itself approximates the 'full' solution, is called a first order layer. Higher order layers are defined accordingly. In the sequel the term 'layer' always refers to a 'zeroth order layer'.
We now collect the results:

- a junction layer of width $O(\lambda|\ln \lambda|)$ occurs at the jump-discontinuity $x = X$ of the reduced solution,
- a boundary layer of the width $O(\lambda|\ln \lambda|)$ occurs at the Dirichlet boundary point $x = 1$ unless the prescribed boundary value is attained by the reduced solution.
- a first order layer occurs at the Neumann boundary point $x = -1$.

By a 'heuristical' extrapolation of these results to the stationary semiconductor device problem we are led to expect layers of width $O(\lambda|\ln \lambda|)$ to occur at abrupt junctions and at Schottky contacts, since the boundary data do not satisfy the zero-space-charge condition there. The reason for the occurrence of layers at semiconductor-oxide interfaces will be explained later on. We do not expect layers to occur at Ohmic contacts, since the boundary data satisfy the zero-space-charge condition there, and at insulating boundary segments, where homogeneous Neumann boundary conditions are prescribed.
This is fully confirmed by physical reasoning. The occurrence of depletion layers at junctions and Schottky contacts and the occurrence of inversion layers at semiconductor oxide interfaces is well-known and physically well understood (see [4.29], [4.3]).

Slow and Fast Scales

To understand mathematically the structure of solutions of the semiconductor device problem, i.e. to make the 'extrapolation' from the model problem (4.2.1), (4.2.2) to the semiconductor device problem rigorous, we need to abandon explicit calculations as performed for the model problem and instead find a constructive method to analyse the structure of solutions.
The clou to this lies in the observation that the solution φ of the model problem varies on two different scales, namely on a slow one outside the layers and on a fast one inside the layers and that the solution is – up to small terms – a sum of functions, each of which varies only on one of these scales. We now pretend not to know the solution of the model problem, and,

instead, try to construct it approximately by only using this structural information.
Clearly, the derivatives of the slowly varying part are of moderate magnitude. Therefore, to obtain this slowly varying part, we neglect the term $\lambda^2\varphi''$ in (4.2.1). We also neglect the boundary conditions (4.2.2), since we expect the slowly varying part only to approximate the solution away from the boundaries and away from the junction. So we obtain the reduced problem for the zeroth order outer solution:

$$0 = a^2\bar{\varphi}(x) - C(x). \tag{4.2.21}$$

The reduced solution is $\bar{\varphi}(x) = C(x)/a^2$.
To derive the fast varying terms we set:

$$\sigma = \varphi - \bar{\varphi} \tag{4.2.22}$$

and obtain from (4.2.1), (4.2.2):

$$\lambda^2\sigma'' = a^2\sigma \tag{4.2.23}$$

$$\sigma(X+) - \sigma(X-) = \bar{\varphi}(X-) - \bar{\varphi}(X+), \qquad \sigma'(X+) = \sigma'(X-) \tag{4.2.24}$$

(since we require $\varphi \in C^1([-1, 1])$ and since $\bar{\varphi}$ is piecewise constant) and the boundary conditions:

$$\sigma(1) = \beta - \bar{\varphi}(1), \qquad \sigma'(-1) = \alpha. \tag{4.2.25}$$

The function σ is composed of inner solutions.
To analyse the structure of the solution within the junction layer about the discontinuity $x = X$ we 'look' into the layer by using a mathematical microscope represented by a stretching of the independent variable:

$$\omega(x, \lambda) = \frac{x - X}{\lambda}. \tag{4.2.26}$$

We obtain from (4.2.23)–(4.2.25):

$$\ddot{\hat{\varphi}}(\omega) = a^2\hat{\varphi}(\omega) \tag{4.2.27}$$

$$\hat{\varphi}(0+) - \hat{\varphi}(0-) = \bar{\varphi}(X-) - \bar{\varphi}(X+), \qquad \dot{\hat{\varphi}}(0+) = \dot{\hat{\varphi}}(0-), \tag{4.2.28}$$

where $\hat{\varphi}(\omega(x, \lambda))$ represents the junction layer inner term. The dots denote differentiation with respect to ω. Since $\hat{\varphi}(\omega(x, \lambda))$ is supposed to approximate σ only within the junction layer, we do not expect it to satisfy the boundary conditions at the points $x = \pm 1$, which are far outside the junction layer. However, we need additional conditions to determine $\hat{\varphi}$ uniquely.
At a fixed point $x \in (X, 1)$ the solution $\varphi(x, \lambda)$ shall be approximated by the outer term $\bar{\varphi}(x)$, if λ is sufficiently small. Disregarding boundary layer terms we want $\varphi(x, \lambda) \approx \bar{\varphi}(x) + \hat{\varphi}(\omega(x, \lambda))$ to satisfy

$$\lim_{\lambda \to 0+} \varphi(x, \lambda) = \lim_{\lambda \to 0+} (\bar{\varphi}(x) + \hat{\varphi}(\omega(x, \lambda))) = \bar{\varphi}(x).$$

Thus, $\lim_{\lambda \to 0+} \hat{\varphi}(\omega) = \lim_{\lambda \to 0+} \hat{\varphi}((x-X)/\lambda) = \lim_{\omega \to \infty} \hat{\varphi}(\omega) = 0$ has to hold and analogously $\lim_{\omega \to -\infty} \hat{\varphi}(\omega) = 0$ follows. The remaining boundary conditions for $\hat{\varphi}$ therefore are

$$\hat{\varphi}(\infty) = \hat{\varphi}(-\infty) = 0. \tag{4.2.29}$$

These so called matching conditions guarantee that the inner approximation $\varphi_{\mathrm{I}} = \bar{\varphi} + \hat{\varphi}$ matches the outer term $\bar{\varphi}$ away from the junction $x = X$. Note that $\hat{\varphi}$ is defined for $\omega \in (-\infty, \infty)$.

The problem (4.2.27)–(4.2.29) is called (zeroth order) junction layer problem and its independent variable ω is called fast independent junction layer variable.

The junction layer problem can easily be solved explicitely and its solution $\hat{\varphi}$ is given by the junction layer term (4.2.8). $\hat{\varphi}$ varies slowly on the ω-scale, but it varies fast on the x-scale.

To obtain the Dirichlet-layer term we stretch the independent variable close to $x = 1$, introduce the fast independent Dirichlet boundary layer variable:

$$\varrho(x, \lambda) = \frac{1-x}{\lambda} \tag{4.2.30}$$

and obtain the Dirichlet-boundary layer term by proceeding as in the junction-layer case:

$$\tilde{\varphi}_+(\varrho) = (\beta - \bar{\varphi}(1))e^{-a\varrho}, \qquad \varrho \geqq 0 \tag{4.2.31}$$

(see (4.2.9)).

Close to the Neumann boundary point $x = -1$ we perform a stretching by introducing the fast independent variable:

$$\gamma(x, \lambda) = \frac{x+1}{\lambda} \tag{4.2.32}$$

and compute the Neumann boundary term:

$$\tilde{\varphi}_-(\gamma) \equiv 0. \tag{4.2.33}$$

Clearly, $\tilde{\varphi}_-$ only agrees to first order with (4.2.9).

Since all layer terms are small outside the corresponding layers (as guaranteed by the matching conditions) they can be superimposed to obtain an approximation of the solution:

$$\varphi(x, \lambda) \approx \bar{\varphi}(x) + \hat{\varphi}\left(\frac{x-X}{\lambda}\right) + \tilde{\varphi}_+\left(\frac{1-x}{\lambda}\right), \tag{4.2.34}$$

which is uniform on the interval $[-1, 1]$. The terms $\bar{\varphi}, \hat{\varphi}, \tilde{\varphi}_+$ do not depend on the perturbation parameter λ explicitely when regarded as functions of the slow and corresponding fast variables resp.

By comparing the right hand side of (4.2.34) with the exact solution (4.2.5) we realize that the approximation error of the matched asymptotic expansion (4.2.34) is $O(\lambda)$ for $\lambda \to 0$.

For variable-coefficent and nonlinear singular perturbation problems, whose solutions are not known a-priori, we need a method to estimate the approximation error. The standard way is to compute or estimate the residual of the expansion, which for our model problem is obtained by inserting the right hand side of (4.2.34) into the problem (4.2.1), (4.2.2). One expects the magnitude of the residual to directly influence the magnitude of the approximation error, in fact, if the singularly perturbed problem is 'uniformly stable' as $\lambda \to 0+$ (see [4.6]) one can prove that the approximation error is bounded by a constant multiple of an appropriate norm of the residual.
A simple computation shows that the residual of (4.2.34) is (as the approximation error) $O(\lambda)$. To obtain an approximation which yields a smaller residual higher order matched asymptotic expansions in the form of series in powers of λ with coefficients, which are sums of slow and fast varying terms, have to be set up.
Asymptotic expansions for solutions of singularly perturbed problems are constructed in such a way that, when cutting the expansion at some index M, then the so obtained finite series satisfies the singularly perturbed problem approximately with an error, which is at least proportional to the $(M+1)$-st power of the perturbation parameter λ. The proportionality constant may of course depend on other possibly occuring problem parameters.
It is a-priori not clear whether there is a solution of the singularly perturbed problem, which is approximated by the finite series for small λ, unless we can prove for the class of problems under investigation that the conclusion from small residual to small approximation error of the solutions is admissible. As long as a result of this type is not available we say that the asymptotic expansion only formally represents a solution of the singularly perturbed problem. If there is a solution of the singularly perturbed problem, which can be approximated to arbitrary accuracy by the 'finite' series, if only the perturbation parameter is small, then we say that the expansion asymptotically represents the solution.
The representation problem is often disregarded in an engineering approach, it is however one of the central mathematical problems of singular perturbation theory. Taking a rigorous point of view, one may only 'accept' an asymptotic expansion, if the representation proof is available. For many real-life problems – like the semiconductor device equations – this is, however, an extraordinary difficult matter. We shall in the sequel take a semi-rigorous point of view and, loosely speaking, 'do what we can' to justify the asymptotic expansions for the semiconductor problem. This means we shall rigorously justify the expansions for models of simple devices and for devices under special biasing conditions and then extend these results by intuitive, physical and numerical reasoning to complicated devices under more general biasing conditions.

4.3. Asymptotic Expansions

Most of the computations and proofs involved in the asymptotic analysis are simplified by using the set of variables (ψ, u, v) instead of (ψ, n, p):

$$n = \delta^2 e^{\psi} u, \qquad p = \delta^2 e^{-\psi} v. \tag{4.3.1}$$

The current densities then read:

$$J_n = \delta^2 \mu_n e^{\psi} \operatorname{grad} u, \qquad J_p = -\delta^2 \mu_p e^{-\psi} \operatorname{grad} v \tag{4.3.2}$$

and the basic semiconductor device equations (4.1.1)–(4.1.5) can be written as singularly perturbed elliptic system with the perturbation parameter λ:

$$\left.\begin{array}{ll} (SP1) & \lambda^2 \Delta \psi = \delta^2 e^{\psi} u - \delta^2 e^{-\psi} v - C(x) \\ (SP2) & \operatorname{div}(\mu_n e^{\psi} \operatorname{grad} u) = S \\ (SP3) & \operatorname{div}(\mu_p e^{-\psi} \operatorname{grad} v) = S \end{array}\right\} \quad x \in \Omega \subseteq \mathbb{R}^k,$$

where we set:

$$S = \frac{1}{\delta^2} R. \tag{4.3.3}$$

In the sequel we shall use the following assumptions:

(A.4.3.1) Ω is a $C^{0,1}$-domain and $\partial\Omega$ is the union of finitely many C^{∞}-segments. There are N junctions in Ω, each of them represented by a $(k-1)$-dimensional C^{∞}-surface $\Gamma_i \subseteq \Omega$, $i = 1, \ldots, N$. The closures of Γ_i are piecewise disjoint, i.e. $\bar{\Gamma}_i \cap \bar{\Gamma}_j = \{\ \}$ for $i \neq j$ and the intersection of $\bar{\Gamma}_i$ with $\partial\Omega$ consists of $(k-2)$-dimensional surfaces for all $i = 1, \ldots, N$ (for $k = 1$ Γ and $\partial\Omega$ are disjoint). The Γ_i's split Ω into $N+1$ disjoint $C^{0,1}$-domains $\Omega_0, \ldots, \Omega_N$: $\Omega = \bigcup_{i=1}^{N} \Gamma_i \cup \bigcup_{i=0}^{N} \Omega_i$.

All junctions are assumed to be abrupt and the doping concentration $|C|$ is bounded below:

(A.4.3.2) $C|_{\bar{\Omega}_i} \in C^{0,1}(\bar{\Omega}_i)$ for $i = 0, \ldots, N$; $[C(x)]_{\Gamma_i} \neq 0$ for all $x \in \Gamma_i$, $i = 1, \ldots, N$ and $|C(x)| \geqq \underset{\sim}{c} > 0$ for all $x \in \Omega - \bigcup_{i=1}^{N} \Gamma_i$.

We call Ω_i an n-domain if

$$C(x) > 0 \qquad \text{for all} \qquad x \in \Omega_i \tag{4.3.4 (a)}$$

and we call it a p-domain if

$$C(x) < 0 \qquad \text{for all} \qquad x \in \Omega_i. \tag{4.3.4 (b)}$$

The geometry-notations are illustrated by the thyristor geometry in Fig. 4.3.1.

We admit position-dependent mobilities, which, because of their possible doping dependence, may have discontinuites at the junctions:

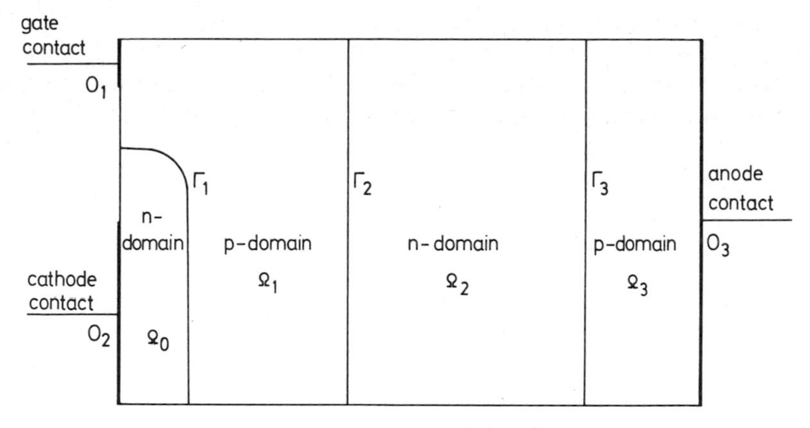

Fig. 4.3.1 Thyristor geometry

(A.4.3.3) (i) $\mu_n = \mu_n(x), \qquad \mu_p = \mu_p(x)$
(ii) $0 < \underline{\mu}_n \leqq \mu_n(x) \leqq \bar{\mu}_n, \qquad 0 < \underline{\mu}_p \leqq \mu_p(x) \leqq \bar{\mu}_p$
(iii) $\mu_n|_{\Omega_i}, \mu_p|_{\bar{\Omega}_i} \in C^{0,1}(\bar{\Omega}_i)$, for $i = 0, \ldots, N$.

The recombination-generation term may be of Shockley-Read-Hall and/or Auger type:

(A.4.3.4) $S = F(x, \psi, u, v)(uv - 1)$, $F: \Omega \times \mathbb{R} \times (0, \infty)^2 \to \mathbb{R}$, $F(x, \psi, u, v) \geqq 0$ for all $(x, \psi, u, v) \in \Omega \times \mathbb{R} \times (0, \infty)^2$,
$F|_{\bar{\Omega}_i \times \mathbb{R} \times (0, \infty)^2} \in C(\bar{\Omega}_i \times \mathbb{R} \times (0, \infty)^2)$ for $i = 0, \ldots, N$ and
$F(x, ., ., .) \in C^1(\mathbb{R} \times (0, \infty)^2)$ uniformly for $x \in \Omega$.

The first step towards a multidimensional singular perturbation analysis is a mathematical formulation of the expected behaviour of the solutions at the critical surfaces. Therefore we proceed as in [4.6], [4.7], [4.10] and introduce local coordinate transformations.

Local Coordinate Transformations

We denote by $r(x)$ the normal distance of a point $x \in \Omega$ to the boundary $\partial\Omega$, i.e.

$$r(x) := \operatorname{dist}(x, \partial\Omega) \geqq 0$$

and by $q(x) = (q_1(x), \ldots, q_k(x))'$ that point on $\partial\Omega$ which is closest to x (the superscript '' denotes transposition). q is uniquely defined almost everywhere in a sufficiently small neighbourhood of $\partial\Omega$ and $\operatorname{grad} r|_{\partial\Omega} = -\nu$ holds almost everywhere in $\partial\Omega$, where we denote by ν the exterior unit normal vector of $\partial\Omega$.

The Jacobian of q

$$\frac{\partial q}{\partial x} := \left(\frac{\partial q_i}{\partial x_j}\right)_{\substack{i = 1, \ldots, k \\ j = 1, \ldots, k}}$$

satisfies $(\partial q/\partial x) \cdot \operatorname{grad} r|_{\partial\Omega} = 0$.
Let Γ stand for one of the device junctions Γ_i and let Γ split Ω into the $C^{0,1}$-subdomains Ω_+ and Ω_-.
We denote by $w(x)$ the 'oriented' normal distance of the point $x \in \Omega$ to the junction Γ, i.e.

$$|w(x)| = \operatorname{dist}(x, \Gamma)$$

and $w(x) > 0$ in Ω_+, $w(x) < 0$ in Ω_-. By $s(x) = (s_1(x), \ldots, s_k(x))'$ we denote that point on Γ, which is closest to x. Clearly w and s are uniquely defined in a sufficiently small neighbourhood of Γ.
$\eta = \operatorname{grad} w|_\Gamma$ is the unit normal vector of Γ pointing into Ω_+ and

$$\frac{\partial s}{\partial x} \cdot \operatorname{grad} w|_\Gamma = 0$$

holds. The local coordinates (r, q) and (w, s) are illustrated in Fig. 4.3.2.
For a function f defined on Ω or $\bar{\Omega}$ we set

$$f^\Gamma(w, s) \equiv f(x), \qquad f^\partial(r, q) \equiv f(x)$$

for those $x \in \Omega(\bar{\Omega})$ for which $(w(x), s(x))$ and $(r(x), q(x))$ are uniquely defined.
For a piecewise continuous function f the one-sided limits satisfy:

$$f^\Gamma(0+, s) := \lim_{w \to 0+} f^\Gamma(w, s) = \lim_{\substack{x \to s \\ x \in \Omega_+}} f(x), \qquad s \in \Gamma$$

$$f^\Gamma(0-, s) := \lim_{w \to 0-} f^\Gamma(w, s) = \lim_{\substack{x \to s \\ x \in \Omega_-}} f(x), \qquad s \in \Gamma.$$

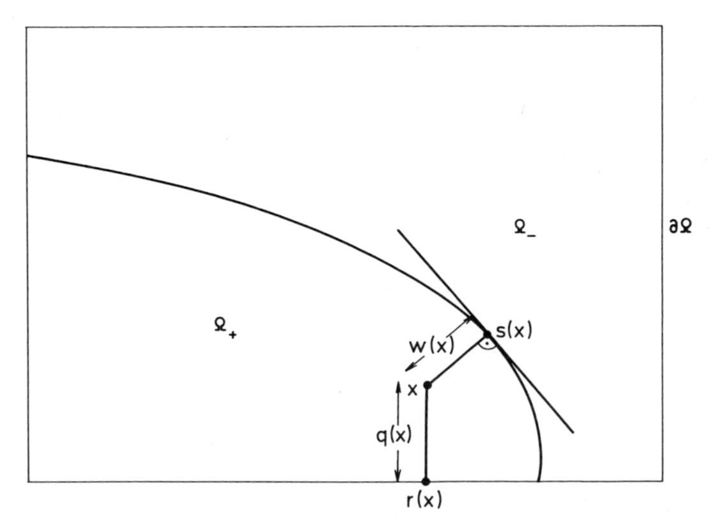

Fig. 4.3.2 Local coordinates

The Outer Expansion

Away from junctions and from the boundary we expect the solutions to be moderately varying functions, i.e. to vary on the x-scale. There we try to represent the solutions by the outer expansion:

$$\begin{pmatrix} \psi(x,\lambda) \\ u(x,\lambda) \\ v(x,\lambda) \end{pmatrix} \sim \begin{pmatrix} \bar{\psi}(x) \\ \bar{u}(x) \\ \bar{v}(x) \end{pmatrix} + \ldots, \tag{4.3.5}$$

where the dots stand for a power series in λ, which starts with the first order term and whose coefficients are functions of x only.
By inserting the expansion into the equations (SP1)–(SP3) and by equating zeroth order terms we obtain the reduced equations:

$$\left.\begin{array}{ll} (RP1) & 0 = \delta^2 e^{\bar{\psi}}\bar{u} - \delta^2 e^{-\bar{\psi}}\bar{v} - C(x) \\ (RP2) & \operatorname{div}(\mu_n e^{\bar{\psi}} \operatorname{grad} \bar{u}) = S(x, \bar{\psi}, \bar{u}, \bar{v}) \\ (RP3) & \operatorname{div}(\mu_p e^{-\bar{\psi}} \operatorname{grad} \bar{v}) = S(x, \bar{\psi}, \bar{u}, \bar{v}) \end{array}\right\} x \in \Omega,$$

which we shall later on supplement by appropriate boundary and interface conditions.
The zeroth order carrier concentration and current density outer terms are derived by inserting the outer expension (4.3.5) into (4.3.1) and (4.3.2) resp.:

$$\bar{n} = \delta^2 e^{\bar{\psi}} u, \qquad \bar{p} = \delta^2 e^{-\bar{\psi}} \bar{v} \tag{4.3.6}$$

$$\bar{J}_n = \delta^2 \mu_n e^{\bar{\psi}} \operatorname{grad} \bar{u}, \qquad \bar{J}_p = -\delta^2 \mu_p e^{-\bar{\psi}} \operatorname{grad} \bar{v}. \tag{4.3.7}$$

The Junction Layer Expansion

Device physics tells us that the direction of steepest decay and growth of the quantities ψ, n and p is orthogonal to device junctions, Schottky contacts and semiconductor-oxide interfaces.
Now let Γ denote one of the junctions Γ_i and $\Omega = \Omega_- \cup \Gamma \cup \Omega_+$. We define the fast independent junction layer variable:

$$\omega(x,\lambda) = \frac{w(x)}{\lambda} \tag{4.3.8}$$

'representing the orthogonal direction to Γ'. The slow variable $s = s(x)$ 'stands for the tangential direction'. In a sufficiently small neighbourhood of Γ we make the following ansatz for the junction layer expansion:

$$\begin{pmatrix} \psi(x,\lambda) \\ u(x,\lambda) \\ v(x,\lambda) \end{pmatrix} \sim \left[\begin{pmatrix} \bar{\psi}(x) \\ \bar{u}(x) \\ \bar{v}(x) \end{pmatrix} + \begin{pmatrix} \hat{\psi}(\omega,s) \\ \hat{u}(\omega,s) \\ \hat{v}(\omega,s) \end{pmatrix}\right] + \lambda\left[\begin{pmatrix} \bar{\psi}_1(x) \\ \bar{u}_1(x) \\ \bar{v}_1(x) \end{pmatrix} + \begin{pmatrix} \hat{\psi}_1(\omega,s) \\ \hat{u}_1(\omega,s) \\ \hat{v}_1(\omega,s) \end{pmatrix}\right] + \ldots. \tag{4.3.9}$$

The dots denote a power series in λ starting with the second order term. Its coefficients are sums of inner and outer terms just as the zeroth and first order coefficients. The matching conditions read:

$$\lim_{\omega \to \pm\infty} \begin{pmatrix} \hat{\psi}(\omega, s) \\ \hat{u}(\omega, s) \\ \hat{v}(\omega, s) \end{pmatrix} = \lim_{\omega \to \pm\infty} \begin{pmatrix} \hat{\psi}_1(\omega, s) \\ \hat{u}_1(\omega, s) \\ \hat{v}_1(\omega, s) \end{pmatrix} = \begin{pmatrix} 0 \\ 0 \\ 0 \end{pmatrix} \quad \text{for all } s \in \Gamma. \tag{4.3.10}$$

We carry out the differentiation in *(SP2)* (assuming that all involved functions are sufficiently smooth away from Γ), insert (4.3.9) and equate coefficients of λ^{-2}. We obtain

$$\hat{u}_{\omega\omega} + \hat{u}_\omega \hat{\psi}_\omega = 0$$

(subscripts 'ω' denote differentiation with respect to ω) and therefore $\hat{u}_\omega = \alpha(s)e^{-\hat{\psi}}$, $\alpha(s) \in \mathbb{R}$. From $\hat{u}(\omega, s) \to 0$ as $\omega \to \pm\infty$ we derive $\hat{u}(\omega, s) \equiv 0$ for all $\omega \in \mathbb{R}$, $s \in \Gamma$. Analogously $\hat{v}(\omega, s) \equiv 0$ follows. The zeroth order u and v-layer terms vanish identically.
Evaluation close to Γ but in Ω_+ gives after equating coefficients of λ^{-1}:

$$\hat{u}_{1\omega\omega} + \hat{\psi}_\omega(\bar{u}_w^\Gamma(0+, s) + \hat{u}_{1\omega}) = 0, \quad \omega > 0, \quad s \in \Gamma$$

(subscripts 'w' denote differentiation with respect to w) and by evaluating in Ω_- we obtain:

$$\hat{u}_{1\omega\omega} + \hat{\psi}_\omega(\bar{u}_w^\Gamma(0-, s) + \hat{u}_{1\omega}) = 0, \quad \omega < 0, \quad s \in \Gamma.$$

These ordinary differential equations are easily integrated and by observing the matching conditions we obtain:

$$\hat{u}_{1\omega}(\omega, s) = \begin{cases} \bar{u}_w^\Gamma(0+, s)\,(e^{-\hat{\psi}(\omega, s)} - 1), & \omega > 0, \quad s \in \Gamma \\ \bar{u}_w^\Gamma(0-, s)\,(e^{-\hat{\psi}(\omega, s)} - 1), & \omega < 0, \quad s \in \Gamma \end{cases}. \tag{4.3.11}$$

Proceeding analogously with *(SP3)* gives:

$$\hat{v}_{1\omega}(\omega, s) = \begin{cases} \bar{v}_w^\Gamma(0+, s)\,(e^{\hat{\psi}(\omega, s)} - 1), & \omega > 0, \quad s \in \Gamma \\ \bar{v}_w^\Gamma(0-, s)\,(e^{\hat{\psi}(\omega, s)} - 1), & \omega < 0, \quad s \in \Gamma \end{cases}. \tag{4.3.12}$$

If $\psi \in L^\infty(\Omega)$, then every weak solution u, v of *(SP2)* and *(SP3)* resp. is in $C(\Omega)$, in particular

$$[u]_\Gamma = [v]_\Gamma = 0$$

holds. We insert the junction layer expansion (4.3.9) into these equations, observe that $\hat{u} \equiv \hat{v} \equiv 0$ holds, equate zeroth order terms and, since this can be done for any junction Γ_i, we obtain the first set of interface conditions for the reduced equations:

(RP4) $[\bar{u}]_{\Gamma_i} = [\bar{v}]_{\Gamma_i} = 0, \quad i = 1, \ldots, N.$

The outer terms $\bar{u}$ and $\bar{v}$ are continuous across each junction.

The internal layer problem is derived by inserting the layer expansion into *(SP1)*, evaluating close to Γ in Ω_+ and Ω_- and by equating zeroth order coefficients:

$$(JLP1)\ \hat{\psi}_{\omega\omega} = \bar{n}^\Gamma(0+,s)e^{\hat{\psi}} - \bar{p}^\Gamma(0+,s)e^{-\hat{\psi}} - C^\Gamma(0+,s), \qquad \omega > 0,\ s \in \Gamma$$
$$(JLP2)\ \hat{\psi}_{\omega\omega} = \bar{n}^\Gamma(0-,s)e^{\hat{\psi}} - \bar{p}^\Gamma(0-,s)e^{-\hat{\psi}} - C^\Gamma(0-,s), \qquad \omega < 0,\ s \in \Gamma.$$

We used (4.3.6) for notational simplicity.
Interface conditions at $\omega = 0$ are obtained by using that the full solution $\psi \in C^1(\Omega)$ (see Section 3.3), which implies $\psi^\Gamma(0+,s) = \psi^\Gamma(0-,s)$ and $\psi^\Gamma_\omega(0+,s) = \psi^\Gamma_\omega(0-,s)$. Inserting the junction layer expansion into these equations and equating zeroth and first order terms resp. gives:

$$(JLP3)\ \hat{\psi}(0+,s) - \hat{\psi}(0-,s) = \bar{\psi}^\Gamma(0-,s) - \bar{\psi}^\Gamma(0+,s)$$
$$(JLP4)\ \hat{\psi}_\omega(0+,s) = \hat{\psi}_\omega(0-,s).$$

To formulate the junction layer problem completely we restate the matching conditions for $\hat{\psi}$:

$$(JLP5)\ \hat{\psi}(\infty,s) = \hat{\psi}(-\infty,s) = 0, \qquad s \in \Gamma.$$

The second set of interface conditions for the reduced problem is derived from the continuity of J_n and J_p across the junction Γ:

$$[\mu_n e^{\psi} \operatorname{grad} u \cdot \eta]_\Gamma = [\mu_p e^{-\psi} \operatorname{grad} v \cdot \eta]_\Gamma = 0. \tag{4.3.13}$$

Since for any sufficiently smooth function f

$$f^\Gamma_w(0,s) = \operatorname{grad} f(s) \cdot \operatorname{grad} w(s), \qquad s \in \Gamma$$

holds, we can rewrite (4.3.13) as

$$\mu^\Gamma_n(0+,s)e^{\psi^\Gamma(0+,s)}u^\Gamma_w(0+,s) = \mu^\Gamma_n(0-,s)e^{\psi^\Gamma(0-,s)}u^\Gamma_w(0-,s), \qquad s \in \Gamma$$
$$\mu^\Gamma_p(0+,s)e^{-\psi^\Gamma(0+,s)}v^\Gamma_w(0+,s) = \mu^\Gamma_p(0-,s)e^{-\psi^\Gamma(0-,s)}v^\Gamma_w(0-,s), \qquad s \in \Gamma$$

and by inserting the expansion (4.3.9) with $\hat{u} \equiv \hat{v} \equiv 0$ we obtain from the first equation:

$$\mu^\Gamma_n(0+,s)e^{\bar{\psi}^\Gamma(0+,s)+\hat{\psi}(0+,s)}(\bar{u}^\Gamma_w(0+,s) + \hat{u}_{1\omega}(0+,s)) =$$
$$= \mu^\Gamma_n(0-,s)e^{\bar{\psi}^\Gamma(0-,s)+\hat{\psi}(0-,s)}(\bar{u}^\Gamma_w(0-,s) + \hat{u}_{1\omega}(0-,s)), \qquad s \in \Gamma.$$

(4.3.11) gives

$$\mu^\Gamma_n(0+,s)e^{\bar{\psi}^\Gamma(0+,s)}\bar{u}^\Gamma_w(0+,s) = \mu^\Gamma_n(0-,s)e^{\bar{\psi}^\Gamma(0-,s)}\bar{u}^\Gamma_w(0-,s), \qquad s \in \Gamma$$

and analogously we derive

$$\mu^\Gamma_p(0+,s)e^{-\bar{\psi}^\Gamma(0+,s)}\bar{v}^\Gamma_w(0+,s) = \mu^\Gamma_p(0-,s)e^{-\bar{\psi}^\Gamma(0-,s)}\bar{v}^\Gamma_w(0-,s),\ s \in \Gamma$$

Thus, the second set of interface conditions for the reduced problem is:

$$(RP5)\ [\mu_n e^{\bar{\psi}} \operatorname{grad} \bar{u} \cdot \eta_i]_{\Gamma_i} = [\mu_p e^{-\bar{\psi}} \operatorname{grad} \bar{v} \cdot \eta_i]_{\Gamma_i} = 0, \qquad i = 1, \ldots, N,$$

where η_i denotes the unit normal vector to the i-th junction Γ_i. The reduced current density-components, which are orthogonal to the junctions, are continuous.

Asymptotic expansions for the carrier concentrations

$$n(x,\lambda) \sim \bar{n}(x)+\hat{n}(\omega,s)+\ldots \quad (4.3.14)\ (a)$$

$$p(x,\lambda) \sim \bar{p}(x)+\hat{p}(\omega,s)+\ldots \quad (4.3.14)\ (b)$$

and for the current densities

$$J_n(x,\lambda) \sim \bar{J}_n(x)+\hat{J}_n(\omega,s)+\ldots \quad (4.3.14)\ (c)$$

$$J_p(x,\lambda) \sim \bar{J}_p(x)+\hat{J}_p(\omega,s)+\ldots, \quad (4.3.14)\ (d)$$

where the dots stand for a power series in λ starting with the first order term, are easily derived from (4.3.1), (4.3.2) and (4.3.9). As mentioned above, the outer terms satisfy (4.3.6), (4.3.7). The n and p-layer terms are given by:

$$\hat{n}(\omega,s) = \begin{cases} \bar{n}^{\Gamma}(0+,s)\,(e^{\hat{\psi}(\omega,s)}-1), & \omega>0, \quad s\in\Gamma \\ \bar{n}^{\Gamma}(0-,s)\,(e^{\hat{\psi}(\omega,s)}-1), & \omega<0, \quad s\in\Gamma \end{cases}. \quad (4.3.15)$$

$$\hat{p}(\omega,s) = \begin{cases} \bar{p}^{\Gamma}(0+,s)\,(e^{-\hat{\psi}(\omega,s)}-1), & \omega>0, \quad s\in\Gamma \\ \bar{p}^{\Gamma}(0-,s)\,(e^{-\hat{\psi}(\omega,s)}-1), & \omega<0, \quad s\in\Gamma \end{cases}. \quad (4.3.16)$$

The current density inner terms are obtained by a simple but lengthy calculation:

$$\hat{J}_n(\omega,s) = \begin{cases} \delta^2\mu_n^{\Gamma}(0+,s)e^{\bar{\psi}^{\Gamma}(0+,s)}\left(\dfrac{\partial s}{\partial x}\right)'\Bigg|_{x=s} \operatorname{grad}_s\bar{u}^{\Gamma}(0+,s)\,(e^{\hat{\psi}(\omega,s)}-1), \\ \qquad \omega>0, \quad s\in\Gamma \\ \delta^2\mu_n^{\Gamma}(0-,s)e^{\bar{\psi}^{\Gamma}(0-,s)}\left(\dfrac{\partial s}{\partial x}\right)'\Bigg|_{x=s} \operatorname{grad}_s\bar{u}^{\Gamma}(0-,s)\,(e^{\hat{\psi}(\omega,s)}-1), \\ \qquad \omega<0, \quad s\in\Gamma \end{cases} \quad (4.3.17)$$

$$\hat{J}_p(\omega,s) = -\begin{cases} \delta^2\mu_p^{\Gamma}(0+,s)e^{-\bar{\psi}^{\Gamma}(0+,s)}\left(\dfrac{\partial s}{\partial x}\right)'\Bigg|_{x=s} \operatorname{grad}_s\bar{v}^{\Gamma}(0+,s)\,(e^{-\hat{\psi}(\omega,s)}-1), \\ \qquad \omega>0, \quad s\in\Gamma \\ \delta^2\mu_p^{\Gamma}(0-,s)e^{-\bar{\psi}^{\Gamma}(0-,s)}\left(\dfrac{\partial s}{\partial x}\right)'\Bigg|_{x=s} \operatorname{grad}_s\bar{v}^{\Gamma}(0-,s)\,(e^{-\hat{\psi}(\omega,s)}-1), \\ \qquad \omega<0, \quad s\in\Gamma. \end{cases} \quad (4.3.18)$$

Since $(\partial s/\partial x)\cdot\eta\,|_{\Gamma}=0$ we conclude

$$\hat{J}_n\cdot\eta|_{\Gamma} \equiv \hat{J}_p\cdot\eta|_{\Gamma} \equiv 0, \quad (4.3.19)$$

i.e. the layer terms of those current density components, which are orthogonal to the junctions, vanish identically while the layer terms of the tangential components may very well be non-zero. The orthogonal current density components are moderately varying and the tangential components may exhibit layer behaviour.

The structure of the current densities is – for $k>1$ – the only real 'multidimensional' feature of the behaviour of solutions at junctions. The structure of solutions in orthogonal direction to the junctions is basically 'dimension-independent', since the 'tangential' coordinate s only appears as parameter in (4.3.15)–(4.3.18). Therefore most of the multidimensional features of the problem are carried by the reduced solutions.

Boundary Layer Expansions

To accurately describe boundary layer phenomena we set up the inner expansion

$$\begin{pmatrix} \psi(x,\lambda) \\ u(x,\lambda) \\ v(x,\lambda) \end{pmatrix} \sim \left[\begin{pmatrix} \bar{\psi}(x) \\ \bar{u}(x) \\ \bar{v}(x) \end{pmatrix} + \begin{pmatrix} \tilde{\psi}(\varrho,q) \\ \tilde{u}(\varrho,q) \\ \tilde{v}(\varrho,q) \end{pmatrix} \right] + \lambda \left[\begin{pmatrix} \bar{\psi}_1(x) \\ \bar{u}_1(x) \\ \bar{v}_1(x) \end{pmatrix} + \begin{pmatrix} \tilde{\psi}_1(\varrho,q) \\ \tilde{u}_1(\varrho,q) \\ \tilde{v}_1(\varrho,q) \end{pmatrix} \right] + \ldots, \tag{4.3.20}$$

where the dots stand for higher order terms,

$$\varrho(x,\lambda) = \frac{r(x)}{\lambda} \tag{4.3.21}$$

is the boundary layer fast independent variable (orthogonal direction) and $q = q(x)$ is the slow independent variable (tangential direction).
The expansions for the carrier concentrations and for the current densities have the same form:

$$\begin{pmatrix} n(x,\lambda) \\ p(x,\lambda) \\ J_n(x,\lambda) \\ J_p(x,\lambda) \end{pmatrix} \sim \left[\begin{pmatrix} \bar{n}(x) \\ \bar{p}(x) \\ \bar{J}_n(x) \\ \bar{J}_p(x) \end{pmatrix} + \begin{pmatrix} \tilde{n}(\varrho,q) \\ \tilde{p}(\varrho,q) \\ \tilde{J}_n(\varrho,q) \\ \tilde{J}_p(\varrho,q) \end{pmatrix} \right] + \ldots . \tag{4.3.22}$$

The matching conditions read:

$$\lim_{\varrho\to\infty} \begin{pmatrix} \tilde{\psi}(\varrho,q) \\ \tilde{u}(\varrho,q) \\ \tilde{v}(\varrho,q) \end{pmatrix} = \lim_{\varrho\to\infty} \begin{pmatrix} \tilde{\psi}_1(\varrho,q) \\ \tilde{u}_1(\varrho,q) \\ \tilde{v}_1(\varrho,q) \end{pmatrix} = \begin{pmatrix} 0 \\ 0 \\ 0 \end{pmatrix}, \quad q \in \partial\Omega \tag{4.3.23 (a)}$$

and

$$\lim_{\varrho\to\infty} \begin{pmatrix} \tilde{n}(\varrho,q) \\ \tilde{p}(\varrho,q) \\ \tilde{J}_n(\varrho,q) \\ \tilde{J}_p(\varrho,q) \end{pmatrix} = \begin{pmatrix} 0 \\ 0 \\ 0 \\ 0 \end{pmatrix}, \qquad q \in \partial\Omega. \tag{4.3.23 (b)}$$

By proceeding as for the junction layer inner expansion we derive

$$\tilde{u}(\varrho,q) \equiv \tilde{v}(\varrho,q) \equiv 0, \qquad \varrho \geqq 0, \qquad q \in \partial\Omega \tag{4.3.24}$$

and

$$\tilde{u}_{1\varrho}(\varrho,q) = \bar{u}_r^\partial(0,q)\,(e^{-\tilde{\psi}(\varrho,q)} - 1), \qquad \varrho \geqq 0, \qquad q \in \partial\Omega, \tag{4.3.25}$$

$$\tilde{v}_{1\varrho}(\varrho,q) = \bar{v}_r^\partial(0,q)\,(e^{\tilde{\psi}(\varrho,q)} - 1), \qquad \varrho \geqq 0, \qquad q \in \partial\Omega, \tag{4.3.26}$$

where the subscripts ϱ and r denote differentiation. u and v do not have zeroth order boundary layers, they vary moderately close to $\partial\Omega$.
In the most general set-up the boundary $\partial\Omega$ is a union of the following disjoint subsets:

$$\partial\Omega = \partial\Omega_D \cup \partial\Omega_N \cup \partial\Omega_I, \qquad \partial\Omega_D = \partial\Omega_O \cup \partial\Omega_S, \tag{4.3.27}$$

where $\partial\Omega_O$ is the union of Ohmic contacts

$$\partial\Omega_O = \bigcup_{i=1}^{r_O} O_i,$$

$\partial\Omega_S$ is the union of Schottky contacts

$$\partial\Omega_S = \bigcup_{j=r_O+1}^{r} S_j,$$

$\partial\Omega_I$ denotes a semiconductor-oxide interface and $\partial\Omega_N$ the union of Neumann segments.
We start with the analysis of the asymptotic behaviour of solutions at Ohmic contacts.

A. Ohmic Contacts

The boundary conditions are given by (4.1.7), (4.1.8). By a simple calculation we obtain the Dirichlet data for n, p and ψ:

$$n|_{\partial\Omega_O} = \frac{1}{2}(C+\sqrt{C^2+4\delta^4})|_{\partial\Omega_O}, \tag{4.3.28 (a)}$$

$$p|_{\partial\Omega_O} = \frac{1}{2}(-C+\sqrt{C^2+4\delta^4})|_{\partial\Omega_O} \tag{4.3.28 (b)}$$

(SP4) (a) $\psi|_{O_i} = \psi_{\mathrm{bi}}|_{O_i} + U_i,$

and from (4.3.1):

(SP4) (b) $u|_{O_i} = e^{-U_i}, \quad v|_{O_i} = e^{U_i}.$

By using that the Dirichlet data satisfy the zero-space-charge condition *(RP1)* it can easily be shown that all zeroth order layer terms vanish at Ohmic contacts:

$$\tilde{\psi}(\varrho, q) \equiv \tilde{u}(\varrho, q) \equiv \tilde{v}(\varrho, q) \equiv 0, \quad \varrho \geqq 0, \quad q \in \partial\Omega_O \tag{4.3.29}$$

$$\tilde{n}(\varrho, q) \equiv \tilde{p}(\varrho, q) \equiv \tilde{J}_n(\varrho, q) \equiv \tilde{J}_p(\varrho, q) \equiv 0, \ \varrho \geqq 0, \ q \in \partial\Omega_O. \tag{4.3.30}$$

Therefore the boundary conditions for the reduced equations at Ohmic contacts are the same as those for the full equations:

(RP6) $\bar{u}|_{O_i} = e^{-U_i}, \quad \bar{v}|_{O_i} = e^{U_i}, \quad \bar{\psi}|_{O_i} = \psi_{\mathrm{bi}}|_{O_i} + U_i.$

Clearly, the outer terms $\bar{n}, \bar{p}$ satisfy (4.1.7), (4.3.28).

B. Neumann Segments

We require the outward electric field and the outward current densities to vanish on $\partial\Omega_N$:

$$\left.\frac{\partial\psi}{\partial\nu}\right|_{\partial\Omega_N} = J_n\cdot\nu|_{\partial\Omega_N} = J_p\cdot\nu|_{\partial\Omega_N} = 0. \tag{4.3.31}$$

The current relations (4.3.2) immediately give homogeneous Neumann boundary conditions for u and v:

$$(SP5)\quad \left.\frac{\partial\psi}{\partial\nu}\right|_{\partial\Omega_N} = \left.\frac{\partial u}{\partial\nu}\right|_{\partial\Omega_N} = \left.\frac{\partial v}{\partial\nu}\right|_{\partial\Omega_N} = 0.$$

By inserting the expansion (4.3.20) and by observing (4.3.24) we obtain:

$$\frac{\partial\bar{u}}{\partial\nu}(q) - \tilde{u}_{1\varrho}(0,q) = 0, \qquad q\in\partial\Omega_N$$

when equating zeroth order coefficients. Since $\frac{\partial\bar{u}}{\partial\nu}(q) = -\bar{u}_r^{\partial}(0,q)$ we derive from (4.3.25):

$$\bar{u}_r^{\partial}(0,q) = 0, \qquad q\in\partial\Omega_N.$$

By proceeding analogously $\bar{v}_r^{\partial}(0,q) = 0$ follows. Thus, the first order u- and v-layer terms $\tilde{u}_1$ and $\tilde{v}_1$ vanish identically, too, and the boundary conditions for the reduced equations at insulating segments are:

$$(RP7)\quad \left.\frac{\partial\bar{u}}{\partial\nu}\right|_{\partial\Omega_N} = \left.\frac{\partial\bar{v}}{\partial\nu}\right|_{\partial\Omega_N} = 0.$$

C. Schottky Contacts

The mixed Schottky contact boundary conditions read:

$$(SP6)\ (a)\quad \psi|_{S_j} = \psi_{bi}|_{S_j} + U_j - \varphi_j$$

$$\left.J_n\cdot\nu\right|_{\partial\mathring{\Omega}_s} = -v_n\left(n - \frac{C+\sqrt{C^2+4\delta^4}}{2}e^{-\varphi_j}\right)\Bigg|_{\partial\mathring{\Omega}_s} \tag{4.3.32}$$

$$\left.J_p\cdot\nu\right|_{\partial\mathring{\Omega}_s} = v_p\left(p - \frac{-C+\sqrt{C^2+4\delta^4}}{2}e^{\varphi_j}\right)\Bigg|_{\partial\mathring{\Omega}_s}. \tag{4.3.33}$$

(see Section 2.3). They transform to:

$$(SP6)\ (b)\left(\delta^2 e^{\psi}\frac{\partial u}{\partial\nu} + v_n\frac{C+\sqrt{C^2+4\delta^4}}{2}e^{U_j-\varphi_j}(u-e^{-U_j})\right)\Bigg|_{\mathring{S}_j} = 0$$

$$(SP6)\ (c)\left(\delta^2 e^{-\psi}\frac{\partial v}{\partial\nu} + v_p\frac{-C+\sqrt{C^2+4\delta^4}}{2}e^{-U_j+\varphi_j}(v-e^{U_j})\right)\Bigg|_{\mathring{S}_j} = 0.$$

By inserting the inner expansion of ψ into *(SP6) (a)* we obtain

$$\bar{\psi}(q)+\tilde{\psi}(0,q) = \psi_{\mathrm{bi}}(q)+U_j-\varphi_j, \quad q\in S_j \tag{4.3.34}$$

and by inserting into *(SP6) (b)*

$$\delta^2 e^{\bar{\psi}(q)+\tilde{\psi}(0,q)}(-\bar{u}_r^{\partial}(0,q)-\tilde{u}_{1\varrho}(0,q))$$
$$+\, v_n \frac{C+\sqrt{C^2+4\delta^4}}{2} e^{U_j-\varphi_j}(\bar{u}(q)-e^{-U_j}) = 0, \quad q\in \mathring{S}_j.$$

(4.3.25) and (4.3.34) give

$$-\delta^2 e^{\bar{\psi}(q)}\bar{u}_r^{\partial}(0,q)+v_n \frac{C+\sqrt{C^2+4\delta^4}}{2} e^{U_j-\varphi_j}(\bar{u}(q)-e^{-U_j}) = 0, \quad q\in \mathring{S}_j.$$

Therefore the conditions *(SP6) (b)* and *(c)* also hold for the reduced equations:

$$(RP8)\ (a)\quad \left.\left(\delta^2 e^{\bar{\psi}}\frac{\partial \bar{u}}{\partial \nu}+v_n \frac{C+\sqrt{C^2+4\delta^4}}{2} e^{U_j-\varphi_j}(\bar{u}-e^{-U_j})\right)\right|_{\mathring{S}_j} = 0$$

$$(RP8)\ (b)\quad \left.\left(\delta^2 e^{-\bar{\psi}}\frac{\partial \bar{v}}{\partial \nu}+v_p \frac{-C+\sqrt{C^2+4\delta^4}}{2} e^{-U_j+\varphi_j}(\bar{v}-e^{U_j})\right)\right|_{\mathring{S}_j} = 0.$$

The layer equation reads:

$(SLP1)$ $\tilde{\psi}_{\varrho\varrho} = \bar{n}(q)e^{\tilde{\psi}}-\bar{p}(q)e^{-\tilde{\psi}}-C(q), \quad \varrho>0, \quad q\in S_j.$

Boundary conditions are derived from (4.3.34), (4.3.23) (a):

$(SLP2)$ $\tilde{\psi}(0,q) = \psi_{\mathrm{bi}}(q)+U_j-\varphi_j-\bar{\psi}(q), \quad q\in S_j$

$(SLP3)$ $\tilde{\psi}(\infty,q) = 0, \quad q\in S_j.$

We shall later on derive the carrier concentration and current-density inner terms.

D. Semiconductor-Oxide Interfaces

A MOS-device is characterized by the occurence of a $(C^{0,1})$-oxide domain Φ adjacent to the semiconductor domain Ω such that $\bar{\Phi}\cap\bar{\Omega} = \overline{\partial\Omega_I}$. Laplace's equation

$(SP7)$ $\Delta\psi = 0, \quad x\in\Phi$

and the interface conditions

$$(SP8)\quad [\psi]_{\partial\Omega_I} = \left[\varepsilon\frac{\partial\psi}{\partial\nu}\right]_{\partial\Omega_I} = 0, \quad \left.\frac{\partial u}{\partial \nu}\right|_{\partial\Omega_I} = \left.\frac{\partial v}{\partial \nu}\right|_{\partial\Omega_I} = 0$$

hold, where ε is defined by

$$\varepsilon(x) = \begin{cases} \varepsilon_s, & x \in \Omega \\ \varepsilon_o, & x \in \Phi \end{cases} \tag{4.3.35}$$

ε_s and ε_o denote the semiconductor and oxide permittivities resp. The oxide boundary $\partial\Phi$ splits into the interface $\partial\Omega_I$, a contact segment $\partial\Phi_c$ on which a Dirichlet condition is imposed:

(SP9) (a) $\psi|_{\partial\Phi_C} = U_G - U_F - \varphi_G =: \psi_G$

and Neumann segments $\partial\Phi_N$:

(SP9) (b) $\left.\dfrac{\partial\psi}{\partial\xi}\right|_{\partial\Phi_N} = 0.$

U_G denotes the scaled externally applied potential, U_F the scaled flat band voltage, φ_G the scaled metal-semiconductor work function difference and ξ the exterior unit normal vector to $\partial\Phi$ ($\xi|_{\partial\Omega_I} = -\nu|_{\partial\Omega_I}$ holds).
The carrier concentrations – and thus u and v – do not exist in the oxide Φ. From *(SP8)* and (4.3.24), (4.3.25), (4.3.26) we immediately obtain the boundary conditions for the reduced equations:

(RP9) $\left.\dfrac{\partial\bar{u}}{\partial\nu}\right|_{\partial\Omega_I} = \left.\dfrac{\partial\bar{v}}{\partial\nu}\right|_{\partial\Omega_I} = 0.$

The boundary layer equation is analogous to *(SLP1)*:

(IP1) $\tilde{\psi}_{\varrho\varrho} = \bar{n}(q)e^{\tilde{\psi}} - \bar{p}(q)e^{-\tilde{\psi}} - C(q), \quad \varrho > 0, \quad q \in \partial\Omega_I$

and the matching condition reads:

(IP2) $\tilde{\psi}(\infty, q) = 0, \quad q \in \partial\Omega_I.$

To derive the boundary condition for $\tilde{\psi}$ at $\varrho = 0$ we assume for simplicity that Φ is a cuboid in $\mathbb{R}^k$, which is aligned to the coordinate system (see Fig. 2.3.2):

$$\Phi = \{x = (x_1, \ldots, x_k) \in \mathbb{R}^k \mid a_i < x_i < b_i\} \tag{4.3.36}$$

and that the scaled width of the oxide layer is $d = b_2 - a_2$. We set $a_2 = 0$, $b_2 = d$. The interface is assumed to be given by

$$\partial\Omega_I = \{x \in \mathbb{R}^k \mid a_i < x_i < b_i, i \neq 2; \quad x_2 = 0\} \tag{4.3.37}$$

and the gate contact $\partial\Phi_c$ is the 'opposite' surface

$$\partial\Phi_c = \{x \in \mathbb{R}^k \mid a_i \leqq x_i \leqq b_i, i \neq 2; \quad x_2 = d\}. \tag{4.3.38}$$

In practice d is of the order of magnitude λ whereas $b_1 - a_1$ and $b_3 - a_3$ are significantly larger. We set

$$d = \alpha\lambda \tag{4.3.39}$$

for some constant $\alpha > 0$ and rescale the independent variable

$$\tilde{x}_i := x_i \quad \text{for} \quad i \neq 2 \quad \text{and} \quad \tilde{x}_2 := \frac{x_2}{d}, \, x \in \Phi \leftrightarrow \tilde{x} \in \tilde{\Phi}. \tag{4.3.40}$$

Then $\gamma(\tilde{x}) := \psi(x)$ satisfies

$$\frac{\partial^2 \gamma}{\partial \tilde{x}_2^2} + d^2 \sum_{i \neq 2} \frac{\partial^2 \gamma}{\partial \tilde{x}_i^2} = 0, \qquad \tilde{x} \in \tilde{\Phi}.$$

(SP8) and *(SP9) (a)* give

$$\gamma|_{\tilde{x}_2 = 1} = \psi_G, \qquad \left.\frac{\partial \gamma}{\partial \tilde{x}_2}\right|_{\tilde{x}_2 = 0} = d \frac{\varepsilon_s}{\varepsilon_o} \left.\frac{\partial \psi}{\partial \nu}\right|_{x_2 = 0}.$$

In MOS-technology it normally happens that the interface is hit by one or more device junctions. For the following we assume that the interface is orthogonal to any such junction Γ, i.e.

$$\operatorname{grad} w(q) \cdot \operatorname{grad} r(q) = 0 \qquad \text{for} \qquad q \in \bar{\Gamma} \cap \overline{\partial \Omega}_I.$$

Then the derivative $\frac{\partial \hat{\psi}}{\partial \nu}(q)$ of the Γ-junction layer term in direction ν at $q \in \bar{\Gamma} \cap \overline{\partial \Omega}_I$ is $O(1)$ as $\lambda \to 0+$ and we obtain

$$\frac{\partial \psi}{\partial \nu}(q) = -\frac{1}{\lambda} \tilde{\psi}_\varrho(0, q) + \ldots,$$

where the dots denote a power series in λ starting with the $O(1)$-term.
By using (4.3.39) we expand the oxide potential γ into a matched asymptotic expansion in λ. The zeroth order term γ_0 satisfies:

$$\frac{\partial^2 \gamma_0}{\partial \tilde{x}_2^2} = 0 \quad \text{in} \quad \tilde{\Phi}, \quad \gamma_0|_{\tilde{x}_2 = 1} = \psi_G, \quad \left.\frac{\partial \gamma_0}{\partial \tilde{x}_2}\right|_{\tilde{x}_2 = 0} = -\beta \tilde{\psi}_\varrho(0, .)|_{\partial \Omega_I},$$

where we denoted

$$\beta = \frac{\varepsilon_s}{\varepsilon_o} \alpha.$$

Thus $\gamma_0|_{\tilde{x}_2 = 0} = \psi_G - \beta \tilde{\psi}_\varrho(0, q)|_{\partial \Omega_I}$ holds and from the continuity of the potential at the interface we obtain:

(IP3) $\quad -\beta \tilde{\psi}_\varrho(0, q) + \tilde{\psi}(0, q) = \psi_G - \bar{\psi}(q), \qquad q \in \partial \Omega_I$

away from points of intersection of $\partial \Omega_I$ with junctions.
We remark that a semiconductor-oxide interface layer occurs because the scaled width d of the oxide domain is of the same order of magnitude as the perturbation parameter λ. Note that d and λ are not intrinsically related.
The carrier concentration boundary layer terms are easily computed from (4.3.20), (4.3.22) by using (4.3.1):

$$\tilde{n}(\varrho, q) = \bar{n}(q)\,(e^{\tilde{\psi}(\varrho,q)} - 1), \qquad \varrho \geqq 0, \quad q \in \partial\Omega_I \cup \partial\Omega_S, \tag{4.3.41}$$

$$\tilde{p}(\varrho, q) = \bar{p}(q)\,(e^{-\tilde{\psi}(\varrho,q)} - 1), \qquad \varrho \geqq 0, \quad q \in \partial\Omega_I \cup \partial\Omega_S. \tag{4.3.42}$$

From (4.3.20), (4.3.22), (4.3.2), (4.3.25), (4.3.26) we obtain the current density layer terms

$$\tilde{J}_n(\varrho, q) = \delta^2 \mu_n(q) e^{\bar{\psi}(q)}\,(\partial q/\partial x)'|_{x=q}\,\mathrm{grad}_q\, \bar{u}^\partial(0, q)\,(e^{\tilde{\psi}(\varrho,q)} - 1),$$

$$\varrho \geqq 0, \quad q \in \Omega_I \cup \partial\Omega_S. \tag{4.3.43}$$

$$\tilde{J}_p(\varrho, q) = -\delta^2 \mu_p(q) e^{-\bar{\psi}(q)}\,(\partial q/\partial x)'|_{x=q}\,\mathrm{grad}_q\, \bar{v}^\partial(0, q)\,(e^{-\tilde{\psi}(\varrho,q)} - 1),$$

$$\varrho \geqq 0, \; q \in \partial\Omega_I \cup \partial\Omega_S. \tag{4.3.44}$$

The layer terms vanish identically on the Ohmic contacts $\partial\Omega_O$ and on the insulating segments $\partial\Omega_N$.

4.4 The Zero-Space-Charge Approximation

The reduced problem, called zero-space-charge approximation in device physics, is completely defined by *(RP1)–(RP9)*.
The reduced potential $\bar{\psi}$ can be computed from *(RP1)* in terms of $\bar{u}$ and $\bar{v}$:

$$\bar{\psi}(\bar{u}, \bar{v})(x) = \ln\left[\frac{C(x) + \sqrt{C(x)^2 + 4\delta^4 \bar{u}\bar{v}}}{2\delta^2 \bar{u}}\right]. \tag{4.4.1}$$

The existence proof for the 'full' problem *(SP1)–(SP9)* as given in Section 3.2 can – after minor modifications (using methods of [4.11]) – also be applied to the reduced problem, if only field-independent mobilities are admitted. The loss of the regularizing property of the inverse of the Laplacian is not severe since $\bar{\psi} : (L^2(\Omega))^2 \to L^2(\Omega)$ is continuous at L^∞-functions $\bar{u}, \bar{v}$. The conditions *(RP4)*, *(RP5)* are the 'natural' interface conditions for the equations *(RP2)*, *(RP3)* since they are already incorporated into the usual weak formulations of *(RP2)*, *(RP3)* and therefore do not require a special treatment.
We only state the existence result:

Theorem 4.4.1: *Let the assumptions (A.4.3.1)–(A.4.3.4) hold. Then the problem (RP2)–(RP9), where $\bar{\psi}$ is defined by (4.4.1) has a weak solution $(\bar{u}^*, \bar{v}^*) \in (H^1(\Omega) \cap L^\infty(\Omega) \cap C(\Omega))^2$, which satisfies*

$$e^{-\bar{U}} \leqq \bar{u}^*(x) \leqq e^{-\underline{U}}, \qquad x \in \Omega \tag{4.4.2 (a)}$$

$$e^{\underline{U}} \leqq \bar{v}^*(x) \leqq e^{\bar{U}}, \qquad x \in \Omega \tag{4.4.2 (b)}$$

with $\bar{U} := \max_{i=1,\ldots,r} U_i$, $\underline{U} := \min_{i=1,\ldots,r} U_i$. *Also*

$$\ln\left[\frac{\underline{C} + \sqrt{\underline{C}^2 + 4\delta^4}}{2\delta^2}\right] + \underline{U} \leqq \bar{\psi}^*(x) \leqq \ln\left[\frac{\bar{C} + \sqrt{\bar{C}^2 + 4\delta^4}}{2\delta^2}\right] + \bar{U}, \; x \in \Omega$$

$$\tag{4.4.2 (c)}$$

holds with $\bar{C} := \sup_{\Omega} C(x)$, $\underline{C} := \inf_{\Omega} C(x)$, and $\bar{\psi}^*|_{\Omega_i} \in C(\bar{\Omega}_i - \partial\Omega)$ *for* $i = 0, \ldots, N$.

We set $|U_{max}| = \bar{U} - \underline{U}$ and, since $\bar{n}\bar{p} = \delta^4 \bar{u}\bar{v}$, obtain the estimate:

$$\delta^4 e^{-|U_{max}|} \leqq \bar{n}^*(x)\bar{p}^*(x) \leqq \delta^4 e^{|U_{max}|}, \qquad x \in \Omega \tag{4.4.3}$$

for the reduced np-product. The reduced carrier concentrations can be expressed in terms of $\bar{u}$, $\bar{v}$ by using (4.3.6) and (4.4.1):

$$\bar{n} = \frac{C + \sqrt{C^2 + 4\delta^4 \bar{u}\bar{v}}}{2}, \qquad \bar{p} = \frac{-C + \sqrt{C^2 + 4\delta^4 \bar{u}\bar{v}}}{2} \tag{4.4.4}$$

and the estimates

$$\frac{\underline{C} + \sqrt{\underline{C}^2 + 4\delta^4 \exp(-|U_{max}|)}}{2} \leqq \bar{n}^*(x)$$

$$\leqq \frac{\bar{C} + \sqrt{\bar{C}^2 + 4\delta^4 \exp(|U_{max}|)}}{2}, \tag{4.4.5 (a)}$$

$$\frac{-\bar{C} + \sqrt{\bar{C}^2 + 4\delta^4 \exp(-|U_{max}|)}}{2} \leqq \bar{p}^*(x)$$

$$\leqq \frac{-\underline{C} + \sqrt{\underline{C}^2 + 4\delta^4 \exp(|U_{max}|)}}{2} \tag{4.4.5 (b)}$$

hold for $x \in \Omega$. Note that (4.4.3), (4.4.5) are only valid for those solutions which satisfy (4.4.2). If the recombination-generation term $S \equiv 0$, then all solutions of the reduced problem satisfy the estimates (4.4.2).
For a device without Schottky contacts and without a semiconductor-oxide interface, i.e. $\partial\Omega = \partial\Omega_N \cup \partial\Omega_O$, the estimates (4.4.2), (4.4.3) also hold for a 'full' solution while only estimates weaker than (4.4.5) were proven for the 'full' carrier concentrations (see Section 3.2). For a device with Schottky contacts or a semiconductor-oxide interface the 'full' potential does not generally satisfy (4.4.2) (c). The drop of the 'full' potential at Schottky contacts and semiconductor-oxide interfaces is not accounted for by (4.4.2) (c) since it occurs within the layers.
By using the estimates for $\bar{u}^*$, $\bar{v}^*$ we also derive from (4.4.4):

$$\bar{n}^*(x) = \begin{cases} C(x) + O(\delta^4 e^{|U_{max}|}) & \text{if} \quad C(x) > 0 \\ O(\delta^4 e^{|U_{max}|}) & \text{if} \quad C(x) < 0 \end{cases} \tag{4.4.6 (a)}$$

$$\bar{p}^*(x) = \begin{cases} O(\delta^4 e^{|U_{max}|}) & \text{if} \quad C(x) > 0 \\ -C(x) + O(\delta^4 e^{|U_{max}|}) & \text{if} \quad C(x) < 0 \end{cases} \tag{4.4.6 (b)}$$

for $\delta^4 e^{|U_{max}|}$ sufficiently small. Clearly $\delta^4 e^{|U_{max}|} \ll 1$ holds, if the maximal applied voltage $|U_{max}|$ is much smaller than the maximal built-in-voltage, i.e. if the low-injection condition

$$|U_{\max}| \ll \sup_{\Omega} \psi_{bi} - \inf_{\Omega} \psi_{bi} \tag{4.4.7}$$

is satisfied. Therefore, in low injection, the reduced electron concentration $\bar{n}^*$ is small in p-regions and close to the doping profile in n-regions while the reduced hole concentration $\bar{p}^*$ is small in n-regions and close to the negative doping profile in p-regions. If the recombination-generation rate is identically zero, then this holds for any reduced electron and hole concentration $\bar{n}$ and $\bar{p}$ resp.

The existence of an *isolated* reduced solution satisfying (4.4.2) can be shown for a one-dimensional *pn*-diode in low injection (see [4.17]). Even when the low injection condition holds, isolatedness of reduced solutions of the multidimensional problem is difficult to prove unless very stringent regularity assumptions are imposed.

4.5 The Layer Problems

The layer problems constitute boundary value problems for ordinary second order differential equations posed on infinite intervals. The independent variables are the corresponding fast layer variables ω and ϱ resp. pointing in direction orthogonal to the critical surfaces and the 'tangential' slow coordinates s, q only appear as parameters in the equations and boundary conditions.

The following Lemma contains a fundamentally important existence and decay result for problems of the form:

$$y''(t) = G(y(t)), \qquad 0 \leqq t < \infty \tag{4.5.1 (a)}$$

$$y(0) = \alpha, \qquad y(\infty) = 0. \tag{4.5.1 (b)}$$

Lemma 4.5.1 (P.Fife, [4.7]): *Let $G : \mathbb{R} \to \mathbb{R}$ be a continuously differentiable function, which satisfies $G(y) < 0$ for $y < 0$, $G(y) > 0$ for $y > 0$ (and therefore $G(0) = 0$ and $G'(0) > 0$).*

Then for every $\alpha \in \mathbb{R}$ there exists a unique monotone solution $y(t)$ of the problem (4.5.1), which together with its derivative $y'(t)$ decays exponentially to zero as $t \to \infty$. For every $\sigma > 0$ this solution satisfies the estimate

$$\frac{1}{D(\sigma)} e^{-(\sqrt{G'(0)}+\sigma)t} \leqq |y(t)| \leqq D(\sigma) e^{-(\sqrt{G'(0)}-\sigma)t} \tag{4.5.2}$$

for some constant $D(\sigma) > 0$.

Proof (see [4.7] for details):
Let $\alpha > 0$. By using y as independent and y' as dependent variable it follows that a monotone solution satisfies

$$t = \int_{y(t)}^{\alpha} \frac{d\tau}{\sqrt{2H(\tau)}}, \qquad H(\tau) := \int_0^{\tau} G(s)\, ds. \tag{4.5.3}$$

Conversely, (4.5.3) defines a unique function $y(t)$, which is a solution of (4.5.1). The exponential decay of y is shown by straightforward estimation of the integral.

For $\alpha = 0$ the unique monotone solution is $y(t) \equiv 0$. For $\alpha < 0$ we change the dependent variable by setting $u = -y$ and obtain the problem

$$u'' = -G(-u), \quad 0 \leqq t < \infty$$
$$u(0) = -\alpha > 0, \quad u(\infty) = 0.$$

Thus, for $\alpha < 0$ the unique monotone solution is given by

$$t = \int_{-y(t)}^{-\alpha} \frac{d\tau}{\sqrt{2H(-\tau)}}. \quad \square \tag{4.5.4}$$

It can also be shown that the solution y is isolated in the sense that linearisations of (4.5.1) at y with exponentially decaying inhomogeneities have unique exponentially decaying solutions.

We shall now prove existence and decay results for the layer problems, which are based on Lemma 4.5.1.

Theorem 4.5.1: *Let (A.4.3.1), (A.4.3.2) hold and assume that a reduced solution* $(\bar{\psi}, \bar{n}, \bar{p})$ *with* $\bar{n}(x) > 0$, $\bar{p}(x) > 0$ *for* $x \in \Omega$ *is given.*

(A) (Junction layer problem) If $C^\Gamma(0+, s) > 0$, $C^\Gamma(0-, s) < 0$, *where* Γ *denotes one of the junctions* Γ_i, *then the problem (JLP1)–(JLP5) has a unique piecewise monotone solution* $\hat{\psi}(., s)$, $s \in \Gamma$, *which satisfies for every* $\sigma > 0$:

$$0 < \hat{\psi}(\omega, s) \leqq C_\sigma \exp\left((1-\sigma)\sqrt{\bar{n}^\Gamma(0-, s) + \bar{p}^\Gamma(0-, s)}\,\omega + D_\sigma \sqrt{|\bar{\psi}^\Gamma(0+, s) - \bar{\psi}^\Gamma(0-, s)|}\right) \tag{4.5.5 (a)}$$

for $\omega < -E_\sigma \sqrt{|\bar{\psi}^\Gamma(0+, s) - \bar{\psi}^\Gamma(0-, s)|}$ *and* $s \in \Gamma$,

$$0 < -\hat{\psi}(\omega, s) \leqq C_\sigma \exp\left((-1+\sigma)\sqrt{\bar{n}^\Gamma(0+, s) + \bar{p}^\Gamma(0+, s)}\,\omega + D_\sigma \sqrt{|\bar{\psi}^\Gamma(0+, s) - \bar{\psi}^\Gamma(0-, s)|}\right) \tag{4.5.5 (b)}$$

for $\omega > E_\sigma \sqrt{|\bar{\psi}^\Gamma(0+, s) - \bar{\psi}^\Gamma(0-, s)|}$ *and* $s \in \Gamma$,

where C_σ, D_σ, $E_\sigma > 0$ *depend on* σ *but not on* $\bar{\psi}^\Gamma(0+, s)$, $\bar{\psi}^\Gamma(0-, s)$.

(B) (Interface layer problem) For $\beta > 0$ *the problem (IP1)–(IP3) has a unique monotone solution* $\tilde{\psi}(., q)$, $q \in \partial\Omega_I$, *which satisfies for every* $\sigma > 0$:

$$|\tilde{\psi}(\varrho, q)| \leqq C_\sigma \exp\left((-1+\sigma)\sqrt{\bar{n}(q) + \bar{p}(q)}\,\varrho + D_\sigma \sqrt{|\tilde{\psi}(0, q)|}\right),$$
$$\varrho > E_\sigma \sqrt{|\tilde{\psi}(0, q)|} \quad \text{and} \quad q \in \partial\Omega_I. \tag{4.5.6}$$

C_σ, D_σ, $E_\sigma > 0$ *depend on* σ *but not on* $\tilde{\psi}(0, q)$.

(C) (Schottky contact layer problem) The problem (SLP1)–(SLP3) has a unique monotone solution $\tilde{\psi}(., q)$, $q \in S_j$ *which satisfies the estimate (4.5.6) for every* $\sigma > 0$, $q \in S_j$.

The layer terms $\tilde{\psi}$ and $\hat{\psi}$ depend continuously on q and s resp., if the reduced solutions $\bar{n}$, $\bar{p}$ and $\bar{\psi}$ and the doping profile C (resp. their one-sided limits at junctions) depend continuously on q and s.

Piecewise monotone means monotone on $(0, \infty)$ and on $(-\infty, 0)$.
Theorem 4.5.1 (A) covers the existence of layer solutions at *pn*-junctions. If the signs of the one-sided limits of the doping profile $C^{\Gamma}(0+, s)$, $C^{\Gamma}(0-, s)$ are equal, i.e. if Γ is an abrupt n^+n or p^+p-junction, then it can easily be shown that the existence assertion remains valid and that $\hat{\psi}$ decays to zero exponentially, too.
Proof (A): For notational simplicity we ignore the dependence of the layer and reduced solutions on the parameter s. We also omit the superscript Γ, e.g. $\bar{n}(0+)$ stands for $\bar{n}^{\Gamma}(0+, s)$.
Piecewise monotonicity implies:

$$\operatorname{sgn} \hat{\psi}(0+) = -\operatorname{sgn} \hat{\psi}(0-)$$

since $\hat{\psi}$ has a jump-discontinuity at $\omega = 0$. $\hat{\psi}(0+) = 0$ gives $\hat{\psi}(\omega) \equiv 0$ on $(0, \infty)$ (see Lemma 4.5.1) and *(JLP4)* implies $\hat{\psi}(0-) = 0$. Therefore $\hat{\psi}(\omega) \equiv 0$ on $(-\infty, 0)$ which contradicts *(JLP3)*. Analogously $\hat{\psi}(0-) \neq 0$ follows. Two possible cases remain:

(i) $\hat{\psi}(0+) > 0$, $\hat{\psi}(0-) < 0$, $\hat{\psi}$ is decreasing on $(-\infty, 0)$ and decreasing on $(0, \infty)$

(ii) $\hat{\psi}(0+) < 0$, $\hat{\psi}(0-) > 0$, $\hat{\psi}$ is increasing on $(-\infty, 0)$ and increasing on $(0, \infty)$.

In the case (i) we conclude from the proof of Lemma 4.5.1 that every piecewise monotone solution satisfies

$$\omega = \int_{\hat{\psi}(\omega)}^{\hat{\psi}(0+)} \frac{ds}{\sqrt{2H_+(s)}}, \tag{4.5.7 (a)}$$

$$H_+(s) = \int_0^s (\bar{n}(0+)e^{\tau} - \bar{p}(0+)e^{-\tau} - C(0+))\, d\tau$$

$$= \bar{n}(0+)(e^s - 1) + \bar{p}(0+)(e^{-s} - 1) - C(0+)s, \quad s > 0$$

$$\omega = -\int_{-\hat{\psi}(\omega)}^{-\hat{\psi}(0-)} \frac{ds}{\sqrt{2H_-(-s)}}, \tag{4.5.7 (b)}$$

$$H_-(s) = \int_0^s (\bar{n}(0-)e^{\tau} - \bar{p}(0-)e^{-\tau} - C(0-))\, d\tau$$

$$= \bar{n}(0-)(e^s - 1) + \bar{p}(0-)(e^{-s} - 1) - C(0-)s, \quad s > 0.$$

Differentiating (4.5.7) (a), (b) and evaluating at $\omega = 0+, 0-$ resp. gives

$$\hat{\psi}_{\omega}(0+) = -\sqrt{2H_+(\hat{\psi}(0+))}, \qquad \hat{\psi}_{\omega}(0-) = -\sqrt{2H_-(\hat{\psi}(0-))}$$

and from *(JLP3)*, *(JLP4)* we obtain the equation

$$H_+(\hat{\psi}(0+)) = H_-(\bar{\psi}(0+) - \bar{\psi}(0-) + \hat{\psi}(0+)),$$

which can be solved uniquely for $\hat{\psi}(0+)$:

$$\hat{\psi}(0+) = \frac{C(0-)\,(\bar{\psi}(0+)-\bar{\psi}(0-))+(\bar{n}(0-)-\bar{n}(0+))+(\bar{p}(0-)-\bar{p}(0+))}{C(0+)-C(0-)}. \tag{4.5.8 (a)}$$

We calculate $\hat{\psi}(0-)$ from *(JLP3)*:

$$\hat{\psi}(0-) = \frac{C(0+)\,(\bar{\psi}(0+)-\bar{\psi}(0-))+(\bar{n}(0-)-\bar{n}(0+))+(\bar{p}(0-)-\bar{p}(0+))}{C(0+)-C(0-)}. \tag{4.5.8 (b)}$$

By applying an analogous argument to $-\hat{\psi}$ we obtain the same expressions for $\hat{\psi}(0+)$, $\hat{\psi}(0-)$ in the case (ii).
Therefore a unique piecewise monotone solution $\hat{\psi}$ exists, iff $\hat{\psi}(0+)$, $\hat{\psi}(0-)$ as given by (4.5.8) are different from zero and have opposite signs. From the interface conditions *(RP6)*, *(RP7)* we derive $\bar{\psi}(0-) < \bar{\psi}(0+)$ since $C(0-) < 0$, $C(0+) > 0$ and

$$\begin{aligned}(\bar{n}(0-)-\bar{n}(0+))+(\bar{p}(0-)-\bar{p}(0+))\\ = \frac{e^{\bar{\psi}(0-)-\bar{\psi}(0+)}-1}{e^{\bar{\psi}(0-)-\bar{\psi}(0+)}+1}\,(C(0-)+C(0+)).\end{aligned}$$

We set $z = \bar{\psi}(0-)-\bar{\psi}(0+)$ and rewrite (4.5.8) (a), (b) as

$$\hat{\psi}(0+) = \frac{h_+(z)}{(e^z+1)\,(C(0+)-C(0-))},$$

$$\hat{\psi}(0-) = \frac{h_-(z)}{(e^z+1)\,(C(0+)-C(0-))}$$

where

$$h_+(z) = C(0+)g_1(z)+C(0-)g_2(z),$$
$$h_-(z) = C(0-)g_1(z)+C(0+)g_2(z)$$

and

$$g_1(z) = e^z-1, \qquad g_2(z) = e^z-1-z(e^z+1).$$

Since $\bar{\psi}(0-) < \bar{\psi}(0+)$ we have $z < 0$. $g_1(z) < 0$ holds and a simple computation shows that $g_2(z) > 0$ for $z < 0$. $C(0+) > 0$, $C(0-) < 0$ imply $h_+(z) < 0$ and $h_-(z) > 0$. Thus

$$\hat{\psi}(0+) < 0, \qquad \hat{\psi}(0-) > 0$$

follows and existence is established.
To prove the estimates (4.5.5) (a), (b) we set $\hat{\varphi} = -\hat{\psi}$ and transform *(JLP1)*:

$$\hat{\varphi}_{\omega\omega}(\omega) = f_+(\hat{\varphi}(\omega)) := \bar{p}(0+)e^{\hat{\varphi}(\omega)}-\bar{n}(0+)e^{-\hat{\varphi}(\omega)}+C(0+), \qquad \omega > 0$$
$$\hat{\varphi}(0+) = -\hat{\psi}(0+) > 0, \qquad \hat{\varphi}(\infty) = 0.$$

It is easy to show that

$$f_+(s) \geqq (\sqrt{f'_+(0)}-\sigma)^2 s, \qquad 0 < s \leqq \sigma$$

holds for σ sufficiently small. The estimate

$$F_+(s) = \int_0^s f_+(u)\,du \geqq \begin{cases} (\sqrt{f'_+(0)}-\sigma)^2 \dfrac{s^2}{2}, & 0 < s \leqq \sigma \\ (\sqrt{f'_+(0)}-\sigma)^2 \dfrac{\sigma^2}{2} + f_+(\sigma)(s-\sigma), & s > \sigma \end{cases}$$

follows. From the proof of Lemma 4.5.1 we conclude

$$\omega = \int_{-\hat{\psi}(\omega)}^{-\hat{\psi}(0+)} \frac{ds}{\sqrt{2F_+(s)}} \leqq \frac{1}{\sqrt{f'_+(0)}-\sigma} \ln \frac{\sigma}{|\hat{\psi}(\omega)|}$$
$$+ \frac{1}{f_+(\sigma)} \sqrt{(\sqrt{f'_+(0)}-\sigma)^2 \frac{\sigma^2}{2} + f_+(\sigma)(|\hat{\psi}(0+)|-\sigma)}$$

for $\omega > 0$ and $|\hat{\psi}(\omega)| < \sigma$. Since $f'_+(0) = \bar{n}(0+)+\bar{p}(0+)$ we obtain

$$|\hat{\psi}(\omega)| \leqq \sigma \exp\left(-(\sqrt{n(0+)+p(0+)}-\sigma)\omega + F_\sigma \sqrt{|\hat{\psi}(0+)|}\right)$$

for $\omega > G_\sigma \sqrt{|\hat{\psi}(0+)|}$, where F_σ and G_σ are independent of $\hat{\psi}(0+)$. Then (4.5.7) (a) implies (4.5.5) (a). (4.5.5) (b) is established analogously.
The proofs of (B) and (C) proceed in a similar fashion. □

The ψ-boundary and internal layer terms as well as their derivatives decay to zero exponentially as their corresponding fast variable tends to $\pm\infty$. Also the internal and boundary layer terms $\hat{n}$, $\hat{p}$, $\hat{J}_n$, $\hat{J}_p$ and $\tilde{n}$, $\tilde{p}$, $\tilde{J}_n$, $\tilde{J}_p$ resp. decay to zero exponentially.
We split the depletion layer, which occurs at the junction Γ, into a portion, which lies in Ω_- and into a portion which lies in Ω_+. Then the width of the Ω_+-portion measured at a point $s \in \Gamma$ in direction orthogonal to Γ (see Fig. 4.5.1) is given by:

$$w_J^+(\lambda, s) = \frac{O(\lambda)}{\sqrt{\bar{n}^\Gamma(0+,s)+\bar{p}^\Gamma(0+,s)}} \left(\ln\frac{1}{\lambda} + \sqrt{|\bar{\psi}^\Gamma(0+,s)-\bar{\psi}^\Gamma(0-,s)|}\right) \tag{4.5.9 (a)}$$

and the width of the Ω_--portion:

$$w_J^-(\lambda, s) = \frac{O(\lambda)}{\sqrt{\bar{n}^\Gamma(0-,s)+\bar{p}^\Gamma(0-,s)}} \left(\ln\frac{1}{\lambda} + \sqrt{|\bar{\psi}^\Gamma(0+,s)-\bar{\psi}^\Gamma(0-,s)|}\right). \tag{4.5.9 (b)}$$

The total width of the layer $w_J(\lambda, s) = w_J^+(\lambda, s) + w_J^-(\lambda, s)$ increases with the square root of the potential drop at the junction.
The width of the boundary layers at the Schottky contacts and at the semiconductor-oxide interface is

$$w_B(\lambda, q) = \frac{O(\lambda)}{\sqrt{\bar{n}(q)+\bar{p}(q)}} \left(\ln\frac{1}{\lambda} + \sqrt{|\bar{\psi}(0,q)|}\right), \tag{4.5.10}$$

measured at $q \in \partial\Omega_S \cup \partial\Omega_I$ in direction orthogonal to the boundary. Again, the width increases with the square root of the potential drop at q.
Note that the local widths of the layers can depend significantly on the position coordinates s and q resp.

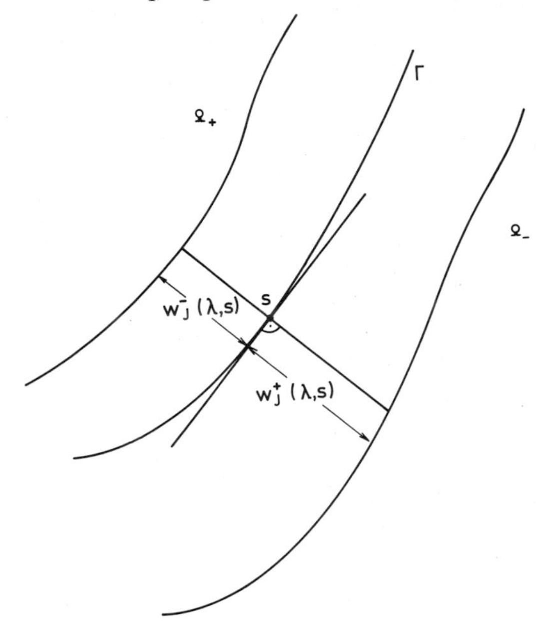

Fig. 4.5.1 Junction layer strip

4.6 Representation and Structure of Solutions

In the preceeding chapters we proved the existence of zeroth order terms of the asymptotic expansions of solutions of the device problem under arbitrary biasing conditions. Thus, we can formally represent the solutions of the device problem in the following form:

$$\psi(x,\lambda) \sim \bar{\psi}(x) + \sum_{i=1}^{N} \hat{\psi}_i\left(\frac{w_i(x)}{\lambda}, s_i(x)\right) + \tilde{\psi}\left(\frac{r(x)}{\lambda}, q(x)\right) + O(\lambda) \quad (4.6.1)\ (a)$$

$$n(x,\lambda) \sim \bar{n}(x) + \sum_{i=1}^{N} \hat{n}_i\left(\frac{w_i(x)}{\lambda}, s_i(x)\right) + \tilde{n}\left(\frac{r(x)}{\lambda}, q(x)\right) + O(\lambda) \quad (4.6.1)\ (b)$$

$$p(x,\lambda) \sim \bar{p}(x) + \sum_{i=1}^{N} \hat{p}_i\left(\frac{w_i(x)}{\lambda}, s_i(x)\right) + \tilde{p}\left(\frac{r(x)}{\lambda}, q(x)\right) + O(\lambda) \quad (4.6.1)\ (c)$$

$$J_n(x,\lambda) \sim \bar{J}_n(x) + \sum_{i=1}^{N} \hat{J}_{ni}\left(\frac{w_i(x)}{\lambda}, s_i(x)\right) + \tilde{J}_n\left(\frac{r(x)}{\lambda}, q(x)\right) + O(\lambda) \quad (4.6.1)\ (d)$$

$$J_p(x,\lambda) \sim \bar{J}_p(x) + \sum_{i=1}^{N} \hat{J}_{pi}\left(\frac{w_i(x)}{\lambda}, s_i(x)\right) + \tilde{J}_p\left(\frac{r(x)}{\lambda}, q(x)\right) + O(\lambda) \quad (4.6.1)\ (e)$$

The functions $\hat{\psi}_i, \hat{n}_i, \ldots, \hat{J}_{pi}$ denote those junction layer terms, which correspond to the i-th device junction.
The expansion procedure can be continued in a straightforward way by expanding the $O(\lambda)$-terms into a matched series and cumbersome but simple calculations lead to problems for inner and outer terms of arbitrary order. The problems defining terms of order larger than zero are linearised versions of the corresponding 'zeroth order problem' (see [4.24]). In the multi-dimensional case we do not have a proof for the isolatedness of reduced solutions, therefore we generally cannot prove the existence of infinite matched asymptotic expansions for the device problem.
For a one-dimensional pn-diode there exists an isolated reduced solution, if the low injection condition (4.4.7) holds (see [4.17]). Since the layer problem is isolatedly soluble, the existence of an infinite expansion follows for this simple device in low injection. Infinite asymptotic expansions also exist for one-dimensional Schottky and MOS-diodes.
The lack of the isolatedness assertion of the reduced solution also extremly complicates the proof of an asymptotic representation result of the 'full' solution by the zeroth order term of the matched expansion (4.6.1) (representation proofs of asymptotic expansions are usually based on the implicit function theorem!). To our knowledge the only result of this type available so far holds for piecewise homogeneously doped devices without Schottky contacts and without a semiconductor-oxide interface in thermal equilibrium. Then the device problem reduces to the equilibrium problem:

$$\lambda^2 \Delta \psi = \delta^2 e^{\psi} - \delta^2 e^{-\psi} - C(x), \qquad x \in \Omega \tag{4.6.2}$$

$$\left.\frac{\partial \psi}{\partial \nu}\right|_{\partial \Omega_N} = 0, \qquad \psi|_{\partial \Omega_O} = \psi_{\text{bi}}|_{\partial \Omega_O}, \tag{4.6.3}$$

where $\partial \Omega = \partial \Omega_N \cup \partial \Omega_O$, $\partial \Omega_O = \bigcup_{i=1}^{r} O_i$. In the proof a few additional geometric assumptions are used, which for simplicity we only formulate for the case of 2 space dimensions ($k = 2$). Also we assume for reasons of notational simplicity that only one device junction Γ occurs:

(A.4.6.1) $\Gamma \subseteq \Omega \subseteq \mathbb{R}^2$ is a C^∞-curve, $\partial \Omega$ a $C^{0,1}$ curve. The intersection of $\bar{\Gamma}$ and $\partial \Omega$ consists of two distinct points S_1, S_2 which have positive distance from $\partial \Omega_O$. Γ splits Ω into the two subdomains Ω_0, Ω_1, i.e. $\Omega = \Omega_0 \cup \Gamma \cup \Omega_1$ holds. There are neighbourhoods of S_1, S_2 within which Γ and $\partial \Omega$ are orthogonal line segments. Moreover the doping profile C is piecewise constant:

$$C(x) = \begin{cases} C_0, & x \in \Omega_0 \\ C_1, & x \in \Omega_1 \end{cases}, \qquad C_0 \neq C_1,\ C_0 \neq 0,\ C_1 \neq 0.$$

The layer solution $\hat{\psi}\left(\frac{w}{\lambda}, s\right)$ is defined in an open strip $S_\chi(\Gamma) = \{x \in \Omega \mid |w(x)| < \chi\}$ about Γ. We choose a C^∞-function θ with the property

$$\theta(w) = \begin{cases} 1, & w \in \left[-\frac{\chi}{2}, \frac{\chi}{2}\right] \\ 0, & |w| \geqq \chi \end{cases}$$

and – facilitated by (A.4.6.1) – extend $\hat{\psi}$ smoothly to Ω by setting

$$\hat{\psi}_\Omega\left(\frac{w(x)}{\lambda}, s(x)\right) := \begin{cases} \hat{\psi}\left(\frac{w(x)}{\lambda}, s(x)\right)\theta(w(x)), & x \in S_\chi(\Gamma) \\ 0, & x \notin S_\chi(\Gamma) \end{cases}. \tag{4.6.4}$$

The following result can be found in [4.18]:

Theorem 4.6.1: *Let (A.4.6.1) hold. Then the unique solution $\psi = \psi_e$ of the equilibrium problem (4.6.2), (4.6.3) satisfies*

$$\psi_e(x) = \psi_{bi}(x) + \hat{\psi}_\Omega\left(\frac{w(x)}{\lambda}, s(x)\right) + O\left(\lambda |\ln \delta|^{\frac{5}{2}}\right) \tag{4.6.5}$$

uniformly for $x \in \Omega$, if λ and $\lambda |\ln \delta|^{\frac{5}{2}}$ are sufficiently small.

Note that the built-in-potential

$$\psi_{bi}(x) = \ln\left[\frac{C(x) + \sqrt{C(x)^2 + 4\delta^4}}{2\delta^2}\right]$$

is the reduced potential $\bar{\psi}$ in thermal equilibrium. The error term in (4.6.5) depends on Ω, C_0 and C_1, it may increase as $\underset{\sim}{c} := \min(|C_0|, |C_1|)$ decreases. Theorem 4.6.1 can easily be extended to multi-junction devices if each junction satisfies the assumptions imposed on Γ in (A.4.6.1) and if the junctions do not intersect.

To illustrate the junction-layer behaviour of solutions of the *pn*-diode problem we present numerical computations for a device, whose geometry is depicted in Fig. 4.6.1. The mobilities were taken constant and equal and the recombination-generation term S was set to zero for the computations.

The 'full' equilibrium potential ψ_e, the equilibrium electron concentration $n_e = \delta^2 e^{\psi_e}$ and the equilibrium hole concentration $p_e = \delta^2 e^{-\psi_e}$ are shown in the Figs. 4.6.2–4.6.4. The *pn*-junction layer is clearly visible. Away from the layers the electron concentration n_e is small in the *p*-region and about equal to the doping profile in the *n*-region. p_e is small in the *n*-region and about equal to the negative doping profile in the *p*-region, just as predicted by (4.4.6) (a) and (b).

Fig. 4.6.5 depicts the potential for -10 Volt (reverse) bias. The width and heigth of the depletion layer increased, but the layer behaviour is still very pronounced.

The Figs. 4.6.6 and 4.6.7 show the electron concentration n and the hole concentration p resp. for 1 Volt (forward) bias. The accumulation of minority carriers, i.e. electrons on the *p*-side and holes on the *n*-side, is clearly visible. Note that for $V = 1$ Volt applied bias we compute $\delta^4 e^{|U|} \approx 2.35 \times 10^3$. Thus we are not dealing with a low injection case. The layer jump and width are

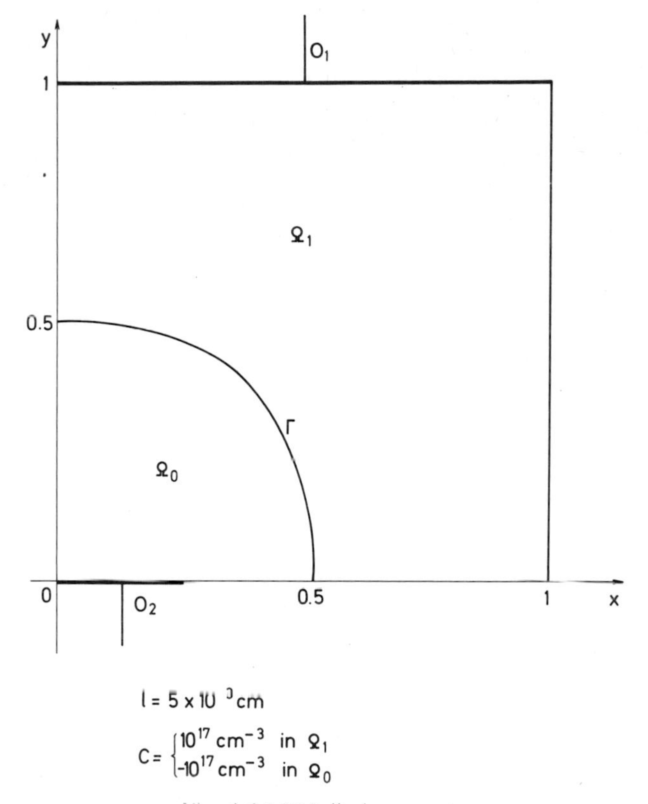

Fig. 4.6.1 PN-diode geometry

visibly smaller than in thermal equilibrium. In Fig. 4.6.8 we show the euclidean norm of the hole current density $|J_p|$. The peak is caused by the singularity of the solution v at the edge of the anode contact (see Section 3.3). The main geometric assumption used to prove the equilibrium representation Theorem 4.6.1 is that the junction Γ and the boundary $\partial\Omega$ are orthogonal line segments locally about their points of intersection.
This makes it possible to smoothly extend the layer solution $\hat{\psi}$ to all of Ω. A priori the layer terms are only defined in that neighbourhood of Γ within which the coordinates $(w(x), s(x))$ are welldefined. Therefore the asymptotic expansion (4.6.1) is locally about those points, at which a junction hits the boundary at an angle different from $\pi/2$, not valid (see Fig. 4.6.9).
We remark that the orthogonality assumption is satisfied by the commonly used simulation geometries.
Also, the expansion fails locally about those points of $\partial\Omega$ at which $\partial\Omega$ is not differentiable, about Schottky contact- and semiconductor-oxide interface edges and locally about the points of intersection of junctions with semiconductor-oxide interfaces (see Fig. 4.6.10). At the latter points the zeroth order term $\bar{\psi}+\hat{\psi}+\tilde{\psi}$ is discontinuous since $\bar{\psi}+\hat{\psi}$ is continuous but $\tilde{\psi}$ is not (because of the condition *(IP3)*, which involves the discontinuous reduced potential $\bar{\psi}$). The local interaction of *pn*-junction and interface layers, which

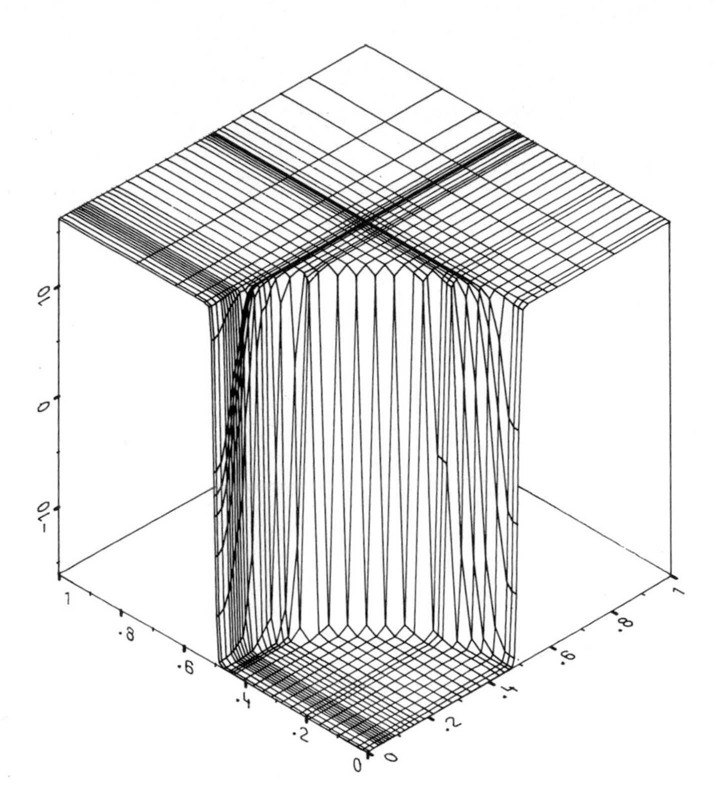

Fig. 4.6.2 Electrostatic potential for 0 V applied bias [4.18]

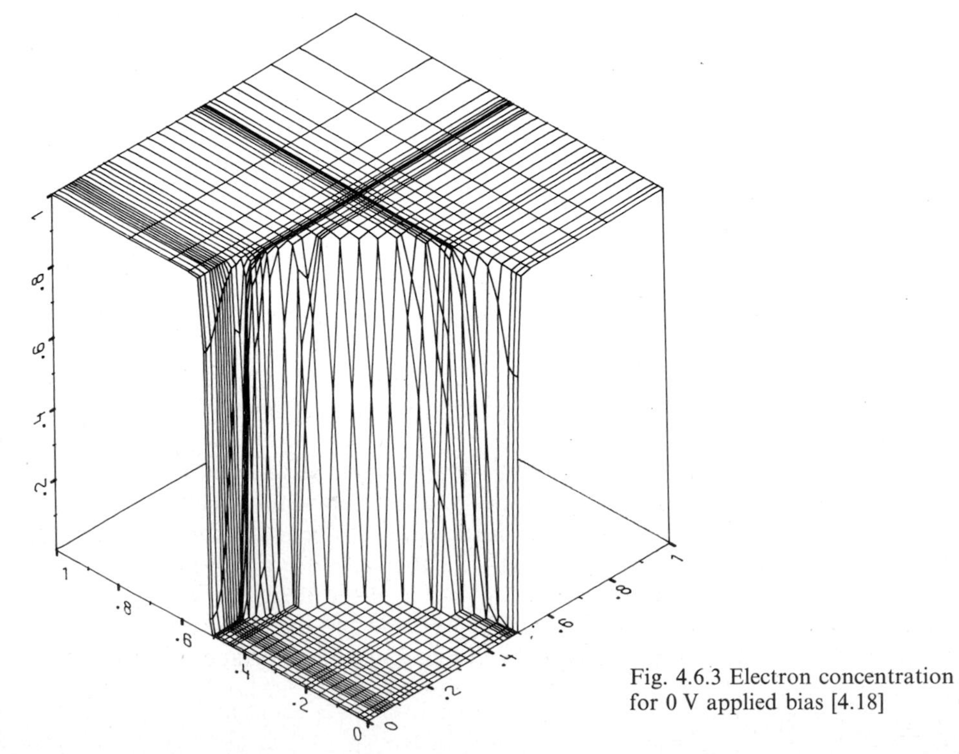

Fig. 4.6.3 Electron concentration for 0 V applied bias [4.18]

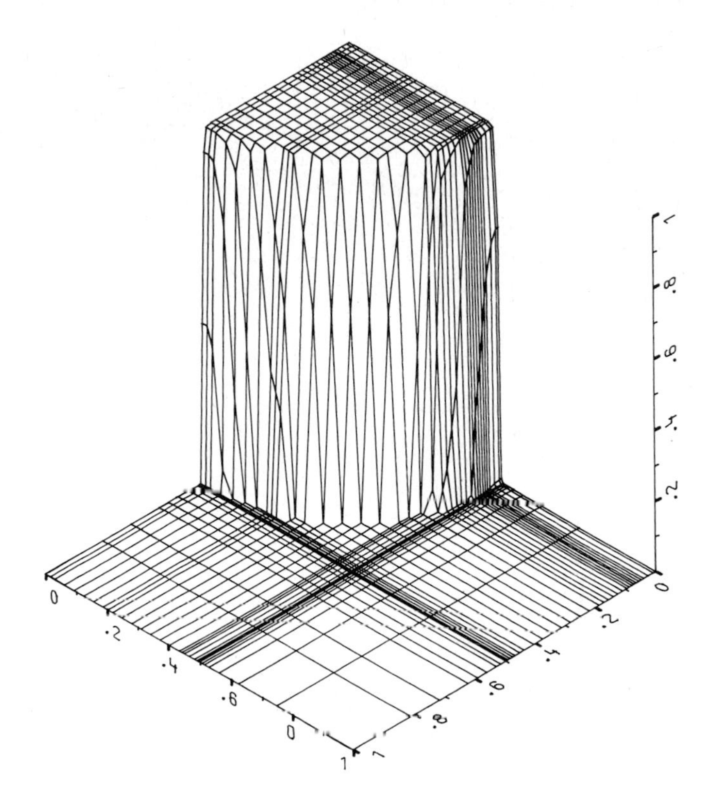

Fig. 4.6.4 Hole concentration for 0 V applied bias [4.18]

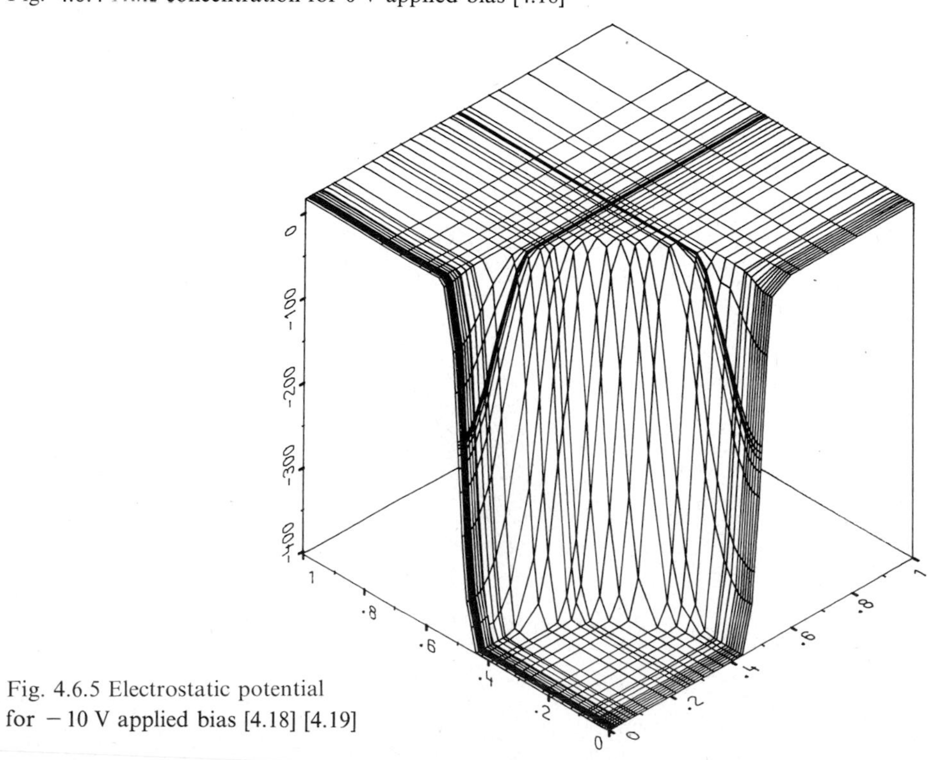

Fig. 4.6.5 Electrostatic potential for − 10 V applied bias [4.18] [4.19]

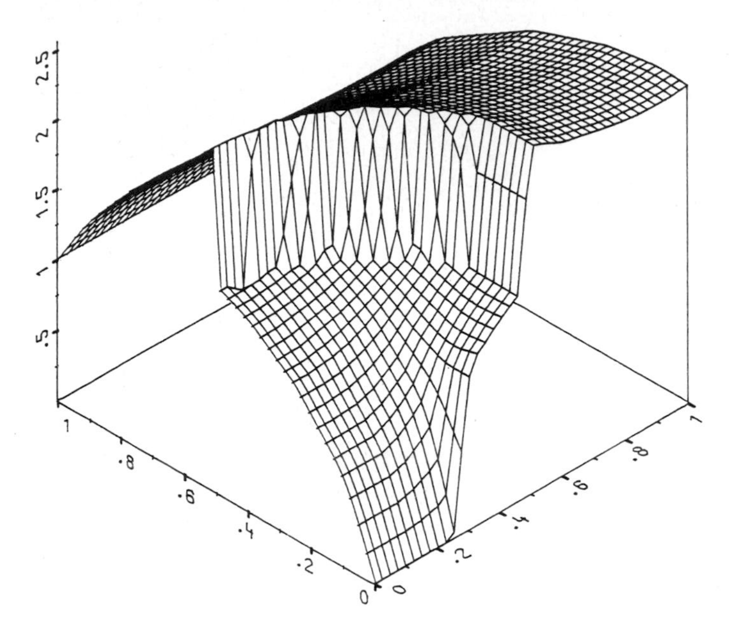

Fig. 4.6.6 Electron concentration for 1 V applied bias

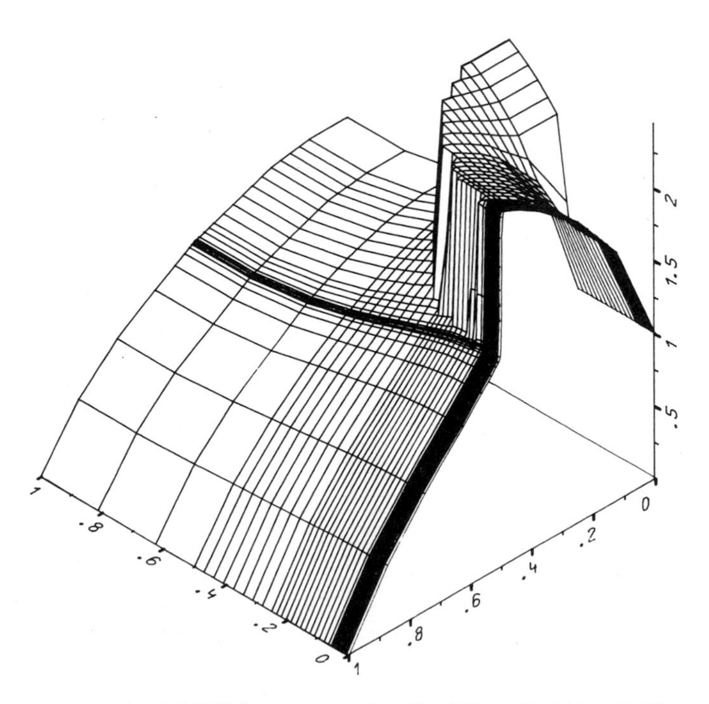

Fig. 4.6.7 Hole concentration for 1 V applied bias [4.18]

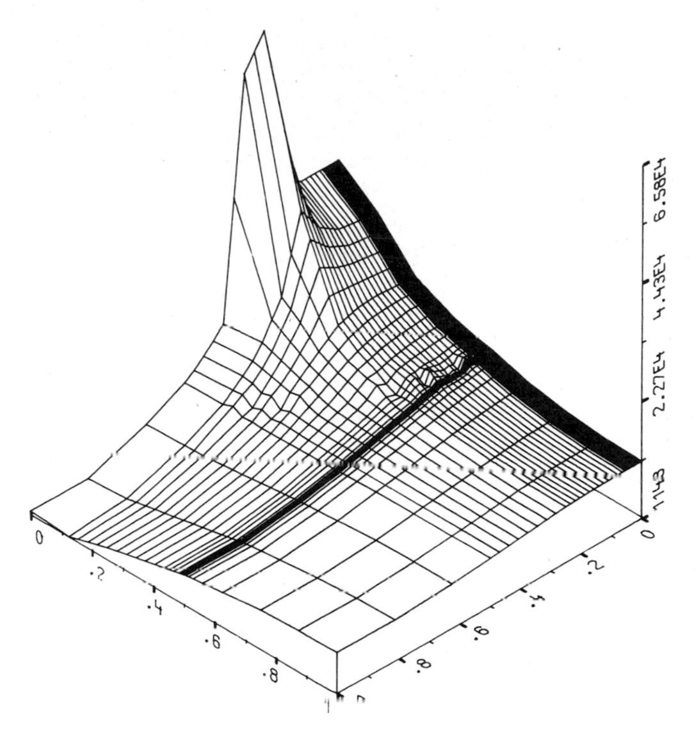

Fig. 4.6.8 Hole current density (modulus) for 1 V applied bias [4.18]

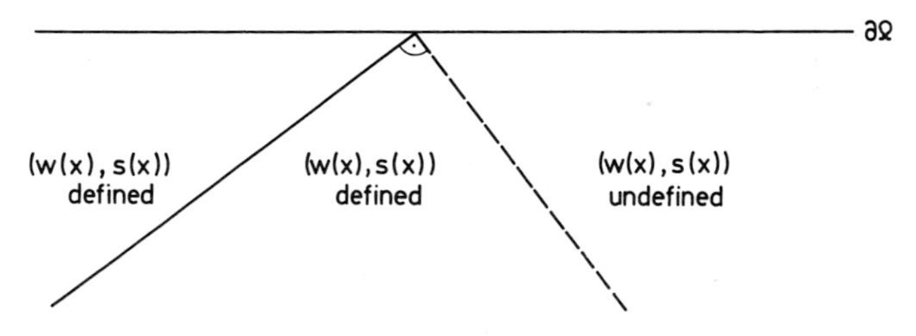

Fig. 4.6.9 Junction-boundary intersection

leads to important physical phenomena (see [4.25], [4.29]), certainly deserves a more intensive mathematical scrutiny than it has been given so far, however, such an analysis would go beyond the scope of this book.
To our knowledge no representation proof for the asymptotic expansions is – even for the simple two-dimensional diode – available under non-equilibrium biasing conditions.
More general representation results exist for certain one-dimensional devices. If the reduced solution is isolated, then it can be shown by using the methods of [4.24] that for given δ and given applied bias U there exist $\lambda_0 = \lambda_0(U, \delta)$ and an isolated solution of the singularly perturbed problem such that the sum of the reduced solution and the layer solution approximates the 'full' solution up to $O(\lambda)$ assuming that $0 < \lambda < \lambda_0$. The coefficient of λ in the error term depends on δ and U. This result is clearly unsatisfactory, too, since it implies that for given maximal doping concentration and applied bias the device length l has to be chosen sufficiently large in order to make λ and the approximation error sufficiently small (see (4.1.6), (4.1.9)). A stronger and practically relevant result was proven in [4.21] under the assumptions of a vanishing recombination-generation rate and constant, equal mobilities for one-dimensional *pn*-diodes with a piecewise homogeneous doping, which is odd about the junction. For these simple devices the approximation error was shown to be reasonably small even for rather large applied bias, if only λ is sufficiently small and δ is not 'too' small compared to λ.
However, for device models, for which $\lambda \ll 1, \lambda \approx \delta$ and which satisfy (A.4.3.1)–(A.4.3.4), the numerical evidence presented above and – for more general geometries – in [4.8], [4.25], [4.27], as well as the analytical results mentioned above for one-dimensional devices clearly demonstrate the following features of the solutions of the stationary device problem under moderate injection:

- There are thin layer strips about abrupt junctions, Schottky contacts and semiconductor-oxide interfaces, whose local width is $O(\lambda(|\ln \lambda| + \sqrt{[\bar{\psi}(x)]}))$, where $[\bar{\psi}(x)]$ denotes the potential drop at the point x on the critical surface. Within these layers ψ, n, p and generally the tangential components of J_n and J_p are rapidly varying functions characterized by the corresponding layer solutions. The electric field at the critical surfaces is of the order of magnitude λ^{-1}.
- Outside the junction-, Schottky contact- and semiconductor-oxide-interface-layers the solutions ψ, n, p, J_n and J_p are moderately varying functions approximated by the reduced solutions $\bar{\psi}$, $\bar{n}$, $\bar{p}$, $\bar{J}_n$ and $\bar{J}_p$ resp.
- Zeroth order layers do not occur at Ohmic contacts and insulating boundary segments.

To illustrate the structure of solutions we depict inner and outer regions for a two-dimensional MOS-transistor in Fig. 4.6.10. *pn*-junction layer strips occur about Γ_1 and Γ_2 and an interface layer about $\partial\Omega_I$.
The singular perturbation approach was applied in [4.3a] to the analysis of the structure of solutions of highly reverse biased *pn*-junction diodes. The authors assumed that the applied bias U satisfies $U = U(\lambda) = -(\text{const}/\lambda^2)$,

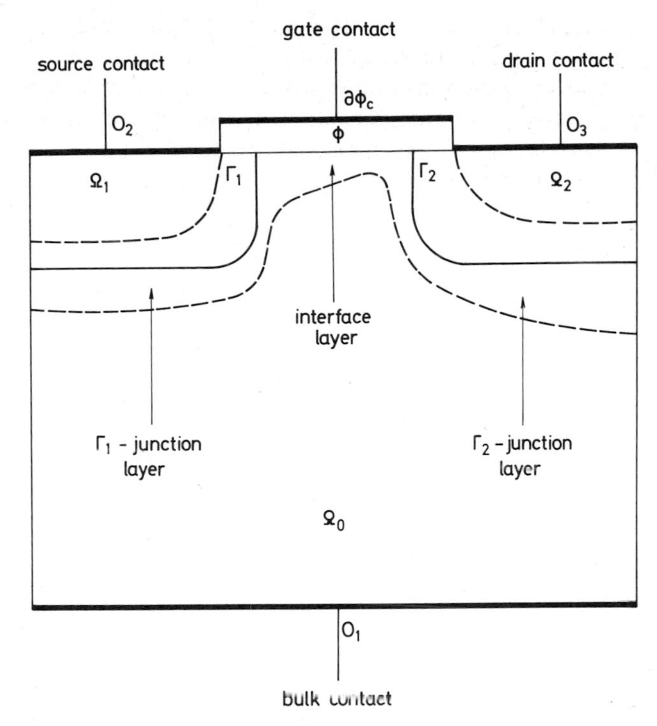

Fig. 4.6.10 Inner and outer regions for a MOS-transistor

they rescaled the potential ψ by U and proved for a one-dimensional, highly simplified case that the solutions of the (rescaled) device problem converge in appropriate norms to the solutions of the reduced problem as $\lambda \to 0+$, $U \to -\infty$. This reduced problem (total depletion approximation) can be rewritten as 'free boundary value problem' and the free boundary represents the edge of the depletion layer. A result of the same type was obtained by Caffarelli and Friedman [4.3b] for multi-dimensional realistic *pn*-diode models.

4.7 Extensions

The most serious shortcomings of the class of device models covered by the presented singular perturbation approach are the requirements that the device junctions are abrupt and the restriction to field-independent mobilities. Also the admitted class of recombination-generation terms does not include the avalanche model for impact ionisation at high electric fields. At this stage, the mathematical understanding of the avalanche phenomenon is very limited, and it is not clear whether and how singular perturbation theory can be applied to give useful qualitative and quantitative results. However, the exclusion of impact ionisation is by no means as serious a shortcoming as those mentioned above, since avalanche generation does not significantly

perturb the device performance at moderate applied voltages, whereas the abrupt-junction assumption and the field-independent mobility requirement have a decisive impact even in low-injection. We shall now demonstrate how these assumptions can be weakened to include physically more relevant doping and mobility models with little extra effort.
Also we shall analyse the impact of lowly doped device regions on the structure of the solutions.

Graded Junctions

Physical doping profiles, obtained by ion-implantation and diffusion (see [4.27]) are not abrupt but continuous, steep, rapidly varying (graded) functions of the space variable x close to device junctions, or, expressed in the terminology of singular perturbation theory, they exhibit internal layer behaviour about junctions (see Fig. 4.7.1). Thus, in the outer regions the doping profile is approximated by a function $\bar{C}$, which varies on the slow scale x, and close to a junction Γ it is represented by $\bar{C}+\hat{C}$, where $\hat{C}$ varies on some fast scale. $\hat{C}$ is small away from the junction. The fast scale is determined by the physical properties of the ion implantation and diffusion processes, however, in order to avoid a two parameter problem, we assume that it is given by $\omega(x, \lambda) = w(x)/\lambda$. There is no intrinsic physical relation between the characteristic Debye length of a device and the width of the doping-layer, but for fixed λ we can adjust the decay behaviour of $\hat{C}$ to the according physical situation. Therefore, without restriction of generality we assume that

$$C(x, \lambda) = \bar{C}(x)+\hat{C}\left(\frac{w(x)}{\lambda}, s(x)\right) \tag{4.7.1}$$

holds close to Γ, where $\bar{C}$ is abrupt:

$$[\bar{C}(x)]_\Gamma \neq 0, \qquad x \in \Gamma. \tag{4.7.2 (a)}$$

The doping layer term satisfies

$$\lim_{\omega \to \pm\infty} \hat{C}(\omega, s) = 0 \qquad \text{for all} \qquad s \in \Gamma. \tag{4.7.2 (b)}$$

The graded profile C is continuous at Γ for every $\lambda > 0$, i.e.:

$$\bar{C}^\Gamma(0+, s)+\hat{C}(0+, s) = \bar{C}^\Gamma(0-, s)+\hat{C}(0-, s)$$
$$\text{for all } s \in \Gamma \tag{4.7.3}$$

holds.
Clearly, this modification does not affect the reduced problem *(RP1)–(RP9)*, it only changes the junction-layer problem. From *(SP1)* we derive

$$\hat{\psi}_{\omega\omega} = \hat{n}-\hat{p}-\hat{C} \tag{4.7.4}$$

and, since (4.3.15), (4.3.16) remain valid, we obtain the junction layer problem for the graded junction:

$$\hat{\psi}_{\omega\omega}(\omega, s) = \bar{n}^{\Gamma}(0+, s)e^{\hat{\psi}(\omega, s)} - \bar{p}^{\Gamma}(0+, s)e^{-\hat{\psi}(\omega, s)}$$
$$-(\bar{C}^{\Gamma}(0+, s) + \hat{C}(\omega, s)),\ \omega > 0,\ s \in \Gamma \tag{4.7.5}$$
$$\hat{\psi}_{\omega\omega}(\omega, s) = \bar{n}^{\Gamma}(0-, s)e^{\hat{\psi}(\omega, s)} - \bar{p}^{\Gamma}(0-, s)e^{-\hat{\psi}(\omega, s)}$$
$$-(\bar{C}^{\Gamma}(0-, s) + \hat{C}(\omega, s)),\ \omega < 0,\ s \in \Gamma \tag{4.7.6}$$
$$\hat{\psi}(0+, s) - \hat{\psi}(0-, s) = \bar{\psi}^{\Gamma}(0-, s) - \bar{\psi}^{\Gamma}(0+, s), \qquad s \in \Gamma \tag{4.7.7}$$
$$\hat{\psi}_{\omega}(0+, s) = \hat{\psi}_{\omega}(0-, s), \qquad s \in \Gamma \tag{4.7.8}$$
$$\hat{\psi}(\infty, s) = \hat{\psi}(-\infty, s) = 0, \qquad s \in \Gamma. \tag{4.7.9}$$

The existence and decay result given in Theorem 4.5.1 (A) does not carry over to the nonautonomous problem (4.7.5)–(4.7.9). An existence proof requires additional assumptions on $\hat{C}$.
Clearly, the decay behaviour of the doping layer term $\hat{C}$ directly influences the decay behaviour of the layer solution $\hat{\psi}$ and – thus – the width of the depletion layer. In most applications the doping profile is modeled as an exponential function close to junctions (see [4.27]), i.e. $\hat{C}$ decays to zero exponentially as $\omega \to \pm\infty$. For these profiles it can easily be shown by using the methods of [4.14], [4.15] that $\hat{\psi}$ decays exponentially as $\omega \to \pm\infty$.
The problem (4.7.5)–(4.7.9) can easily be solved numerically by cutting the infinite interval at 'far out' points X_F and $-X_F$ (see [4.15]) and by using a standard two-point-boundary-value-problem-solver (see, e.g., [4.1], [4.2])

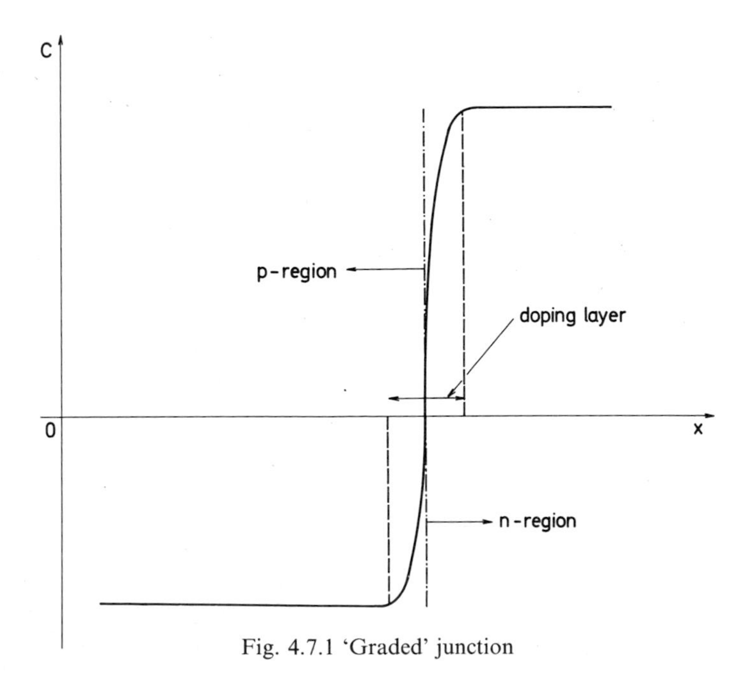

Fig. 4.7.1 'Graded' junction

Strongly One-sided Junctions

In most devices there are junctions, which separate highly doped and lowly doped device regions. An example of a device with a one-sided junction is provided by the asymmetric abrupt *pn*-diode depicted in Fig. 4.7.2.

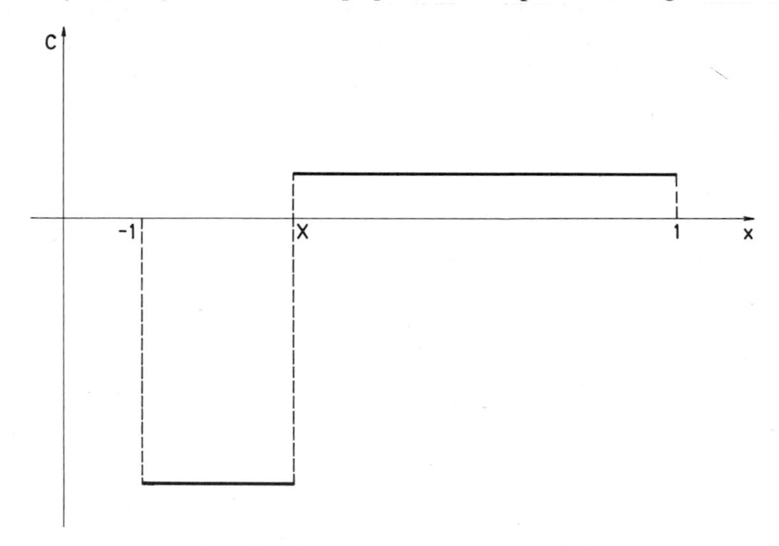

Fig. 4.7.2 'Strongly asymmetric' *PN*-diode

Typically, the doping concentration in the highly doped *p*-region is of the order of magnitude 10^{20} cm^{-3}, while it is 10^{15} cm^{-3} in the lowly doped *n*-region. When the maximal doping concentration C_{max} is used as scaling factor for the carrier concentrations and for the doping profile, then the scaled doping profile is given by:

$$C(x) = \begin{cases} -1, & -1 \leqq x < X \\ c, & X < x \leqq 1 \end{cases} \tag{4.7.10}$$

with $c \ll 1$. Assuming that $\delta^2 = n_i/C_{max}$ is much smaller than c we obtain for the reduced solutions in thermal equilibrium:

$$\bar{n}(x) = \begin{cases} O(\delta^4), & -1 \leqq x < X \\ c + O\left(\dfrac{\delta^4}{c}\right), & X < x \leqq 1 \end{cases}, \tag{4.7.11 (a)}$$

$$\bar{p}(x) = \begin{cases} 1 + O(\delta^4), & -1 \leqq x < X \\ O\left(\dfrac{\delta^4}{c}\right), & X < x \leqq 1 \end{cases} \tag{4.7.11 (b)}$$

and the estimates (4.5.9) (a) for the layer width give:

$$w_J^+ = O\left(\frac{\lambda}{\sqrt{c}}\right)\left(\ln\frac{1}{\lambda} + \sqrt{\psi_{bi}(1) - \psi_{bi}(-1)}\right) \tag{4.7.12 (a)}$$

$$w_J^- = O(\lambda)\left(\ln\frac{1}{\lambda} + \sqrt{\psi_{bi}(1) - \psi_{bi}(-1)}\right), \tag{4.7.12 (b)}$$

where $w_J^+(w_J^-)$ is the width of the layer portion in $[X, 1]$ ($[-1, X]$). By deriving a lower bound of the layer solution $\hat{\psi}$ (using Lemma 4.5.1) it can easily be shown that the estimates (4.7.12) are sharp. Therefore the layer portion located in the lowly doped device region is significantly larger (by a factor of $1/\sqrt{c}$) than the layer portion located in the highly doped device region, if the device is in thermal equilibrium (see Fig. 4.7.3). This assertion can easily be generalized to multidimensional devices under low-injection biasing conditions.

For a silicon device with the characteristic length $1 = 5 \times 10^{-5}$ cm and with the doping profile depicted in Fig. 4.7.2 we compute (for room temperature) $\lambda \approx \delta \approx 10^{-5}$, $c = 10^{-5}$ and thus the width of the layer in the p-region is of the order of magnitude 10^{-4} while the width of the layer in the n-region is of the order magnitude 3×10^{-2} (in thermal equilibrium). Accurately speaking, the potential (and consequently the carrier concentrations) vary in that layer portion, which is located in the lowly doped region, on the scale $(\sqrt{c}/\lambda)x$, which is significantly slower than the scale x/λ determining the variation in the layer portion located in the highly doped device region. Since

$$\frac{\lambda}{\sqrt{c}} = \sqrt{\frac{\varepsilon_s U_T}{l^2 q C_{\min}}}, \qquad C_{\min} := \min_{x \in \Omega} |C(x)| \tag{4.7.13}$$

holds, we conclude that $\lambda_1 := \lambda/\sqrt{c}$ is the local normed Debye length of the lowly doped n-region.

Assume now that the device under consideration has N steeply graded or abrupt junctions Γ_i, $i = 1, \ldots, N$, which split the device domain Ω into $N+1$

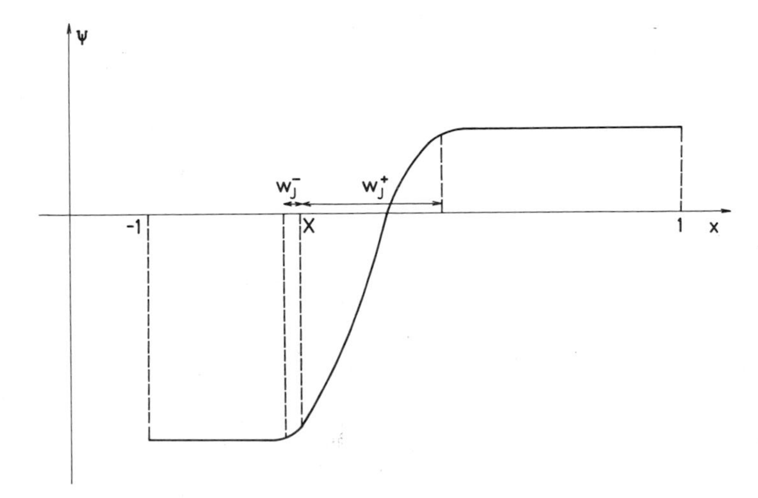

Fig. 4.7.3 Layer structure for one-sided junctions

subdomains Ω_i, $i = 0, \ldots, N$. Then we define the Ω_i-local characteristic normed Debye length by

$$\lambda_i := \sqrt{\frac{\varepsilon_s U_T}{l^2 q \tilde{C}_i}}, \tag{4.7.14}$$

where $\tilde{C}_i$ is of the same order of magnitude as $\max_{x \in \Omega_i} |C(x)|$.
We conclude that – at least in low injection – the solutions of the semiconductor device problem vary within those parts of the layer strips, which are located in the subdomain Ω_i, on the scale determined by the Ω_i-local normed Debye length λ_i.
Generally, the solutions vary faster within the layer strip portions located in the highly doped subdomains than in the portions located in lowly doped subdomains. We cannot – even in thermal equilibrium – expect the asymptotic expansions derived in the previous sections to 'represent' a solution of the device problem, if one local characteristic normed Debye length λ_i is not small. This is well demonstrated by the (hypothetical) *pn*-diode with the (unscaled) doping profile:

$$C(x) = \begin{cases} -10^{20}\,\text{cm}^{-3}, & -l \leqq x < 0 \\ 10^{10}\,\text{cm}^{-3}, & 0 < x \leqq l \end{cases}, \qquad l = 5 \times 10^{-3}\,\text{cm}.$$

The usual scaling gives $\lambda^2 \approx \delta^2 \approx C_{\min}/C_{\max} \approx 10^{-10}$ at room temperature and the equilibrium problem can be written in the form:

$$\lambda^2 \psi'' = \delta^2 e^{\psi} - \delta^2 e^{-\psi} + 1, \qquad -1 \leqq x < 0 \tag{4.7.15 (a)}$$

$$\psi'' = e^{\psi} - e^{-\psi} - 1, \qquad 0 < x \leqq 1 \tag{4.7.15 (b)}$$

$$\psi(\pm 1) = \psi_{\text{bi}}(\pm 1), \; \psi^{(i)}(0+) = \psi^{(i)}(0-) \text{ for } i = 0, 1. \tag{4.7.15 (c)}$$

The solution $\bar{\psi}$ of the zero space charge approximation

$$0 = \delta^2 e^{\bar{\psi}} - \delta^2 e^{-\bar{\psi}} - \begin{cases} -1, & -1 \leqq x < 0 \\ \delta^2, & 0 < x \leqq 1 \end{cases}$$

does not approximate the solution ψ of (4.7.15) well for reasonably small λ in any closed subinterval of (0, 1] since the layer degenerated to the whole *n*-side [0, 1]. The reason for this is that the local normed Debye length λ_1 of the lowly doped *n*-side is of the order of magnitude 1.
These considerations suggest to apply domainwise scaling to the device equations, if strongly one-sided junctions occur. This gives a multi-parameter singular perturbation problem (in the parameters $\lambda_0, \ldots, \lambda_N$). Multi-parameter asymptotic expansions might yield a more accurate description of the structure of solutions than the single-parameter asymptotics, however, rigorous results are not available yet.
We conjecture that the approximation error of the sum of the reduced and layer solutions is generally only $O(\lambda_{\min})$, where $\lambda_{\min} = \min_i \lambda_i$.

Field-dependent Mobilities – Velocity Saturation

We shall now demonstrate that the singular perturbation approach can be extended to cover device models with field-dependent mobilities. We shall here only deal with a special class of mobility models, namely those, which admit velocity saturation at high electric fields (see [4.27]).
Not much attention has been paid so far in the mathematical literature to the phenomenon of velocity saturation, only recently an analysis, which applies to one-dimensional devices was presented in [4.22]. To show the main difficulties and implications of the singular perturbation analysis of velocity saturation device models we analyse symmetric *pn*-junction diodes ($\Omega = (-1, 1)$, $C(x)$ is odd about the junction $X = 0$ and the applied anode potential U_A equals the negative applied cathode potential U_C, i.e. $U_A = -U_C = U/2$).
We employ the mobility model:

$$\mu_n = \mu_p = \mu, \qquad \mu = \mu(\psi') = \frac{v_s}{v_s + |\psi'|}, \qquad v_s > 0. \tag{4.7.16}$$

For more generality and detail we refer the reader to [4.22].
The moduli of the scaled drift velocities v_n^d and v_p^d are given by:

$$|v_n^d| = |\psi'|\mu_n(\psi'), \qquad |v_p^d| = |\psi'|\mu_p(\psi'). \qquad (4.7.17)\ (a)$$

Thus, for the model (4.7.16) we have

$$\lim_{|\psi'|\to\infty} |v_n^d| = \lim_{|\psi'|\to\infty} |v_p^d| = v_s. \qquad (4.7.17)\ (b)$$

v_s is the saturation velocity at high fields. In Fig. 4.7.4 we depict the drift velocities for $\mu_n = \mu_p = 1$ and for (4.7.16).

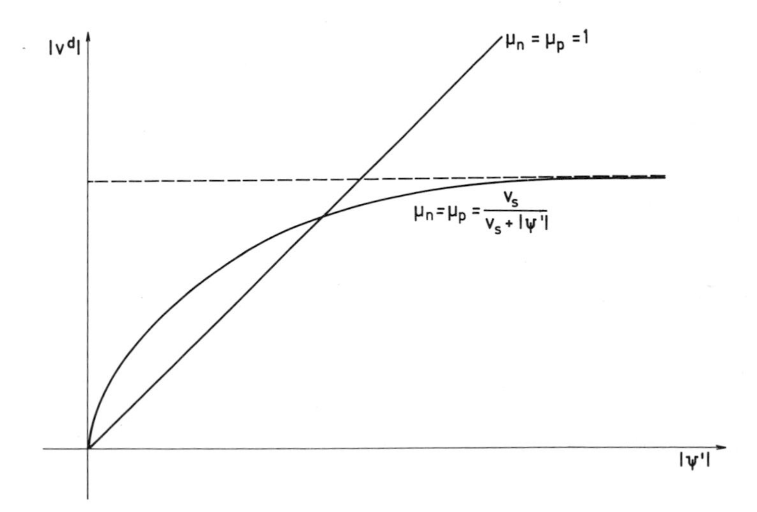

Fig. 4.7.4 Moduli of scaled drift velocities

Motivated by the symmetry of the device we look for solutions, which satisfy

$$\psi(x) = -\psi(-x), \quad n(x) = p(-x), \quad J_n(x) = J_p(-x). \tag{4.7.18}$$

Neglecting recombination-generation effects and assuming a piecewise homogeneous doping we can then write the model for the symmetric *pn*-diode as:

$$\left.\begin{aligned} \lambda^2\psi'' &= n-p-1 && \text{(4.7.19) (a)}\\ n' &= n\psi' + \frac{J}{2\mu(\psi')} && \text{(4.7.19) (b)}\\ p' &= -p\psi' - \frac{J}{2\mu(\psi')} && \text{(4.7.19) (c)}\\ J' &= 0 && \text{(4.7.19) (d)} \end{aligned}\right\} \quad 0 \leqq x \leqq 1,$$

where the x-interval $[0, 1]$ represents the n-side of the device. The boundary conditions read:

$$\psi(1) = \psi_{\text{bi}}(1) - \frac{U}{2} =: \psi_+ \tag{4.7.19) (e}$$

$$n(1) = \frac{1}{2}(1+\sqrt{1+4\delta^4}) =: n_+, \tag{4.7.19) (f}$$

$$p(1) = \frac{1}{2}(-1+\sqrt{1+4\delta^4}) =: p_+ \tag{4.7.19) (g}$$

$$n(0) = p(0), \quad \psi(0) = 0. \tag{4.7.19) (h}$$

We obtain the reduced system of equations by setting λ equal to zero in (4.7.19) (a). With (4.7.16) it reads:

$$\left.\begin{aligned} 0 &= \bar{n} - \bar{p} - 1 && \text{(4.7.20) (a)}\\ \bar{n}' &= \bar{n}\bar{\psi}' + \frac{\bar{J}}{2}\left(\frac{|\bar{\psi}'|}{v_s} + 1\right) && \text{(4.7.20) (b)}\\ \bar{p}' &= -\bar{p}\bar{\psi}' - \frac{\bar{J}}{2}\left(\frac{|\bar{\psi}'|}{v_s} + 1\right) && \text{(4.7.20) (c)}\\ \bar{J}' &= 0 && \text{(4.7.20) (d)} \end{aligned}\right\} \quad 0 \leqq x \leqq 1.$$

Since a zeroth order boundary layer does not occur at the Ohmic contact $x = 1$, we have

$$\bar{n}(1) = n_+, \quad \bar{p}(1) = p_+, \quad \bar{\psi}(1) = \psi_+. \tag{4.7.20) (e}$$

Clearly, we only admit positive reduced electron and hole concentrations $\bar{n}$ and $\bar{p}$ resp.

The junction layer equations are obtained by inserting the matched expansion

$$\begin{pmatrix} \psi(x,\lambda) \\ n(x,\lambda) \\ p(x,\lambda) \\ J(x,\lambda) \end{pmatrix} \sim \begin{pmatrix} \bar{\psi}(x) \\ \bar{n}(x) \\ \bar{p}(x) \\ \bar{J} \end{pmatrix} + \begin{pmatrix} \hat{\psi}(\omega) \\ \hat{n}(\omega) \\ \hat{p}(\omega) \\ 0 \end{pmatrix} + \ldots, \qquad \omega = \frac{x}{\lambda}, \tag{4.7.21}$$

$$\lim_{\omega\to\infty} \hat{\psi}(\omega) = \lim_{\omega\to\infty} \hat{n}(\omega) = \lim_{\omega\to\infty} \hat{p}(\omega) = 0 \tag{4.7.22}$$

into (4.7.19) (a)–(d), multiplying (4.7.19) (b), (c) by λ and by equating zeroth order coefficients:

$$\left.\begin{aligned} \ddot{\hat{\psi}} &= \hat{n} - \hat{p} && \text{(4.7.23) (a)} \\ \dot{\hat{n}} &= \left(\hat{n} + \bar{n}(0) + \frac{\bar{J}}{2v_s}\operatorname{sgn}(\dot{\hat{\psi}})\right)\dot{\hat{\psi}} && \text{(4.7.23) (b)} \\ \dot{\hat{p}} &= -\left(\hat{p} + \bar{p}(0) + \frac{\bar{J}}{2v_s}\operatorname{sgn}(\dot{\hat{\psi}})\right)\dot{\hat{\psi}} && \text{(4.7.23) (c)} \end{aligned}\right\} \quad \omega > 0$$

(the dots denote differentiation with respect to ω). By restraining the investigation to monotone layer potentials $\hat{\psi}$ we have

$$\operatorname{sgn}(\dot{\hat{\psi}}(\omega)) \equiv -\operatorname{sgn}(\hat{\psi}(\omega)) \equiv \operatorname{sgn}(\bar{\psi}(0))$$

since $\bar{\psi}(0) = -\hat{\psi}(0)$. Integration of (4.7.23) (b), (c) gives:

$$\hat{n}(\omega) = \left(\bar{n}(0) + \frac{\bar{J}}{2v_s}\operatorname{sgn}(\bar{\psi}(0))\right)(e^{\hat{\psi}(\omega)} - 1), \qquad \omega > 0 \tag{4.7.24 (a)}$$

$$\hat{p}(\omega) = \left(\bar{p}(0) + \frac{\bar{J}}{2v_s}\operatorname{sgn}(\bar{\psi}(0))\right)(e^{-\hat{\psi}(\omega)} - 1), \qquad \omega > 0 \tag{4.7.24 (b)}$$

From the condition $n(0) = p(0)$ we derive:

$$\bar{n}(0) + \hat{n}(0) = \bar{p}(0) + \hat{p}(0)$$

and by using (4.7.24) and $\hat{\psi}(0) = -\bar{\psi}(0)$:

$$\bar{\psi}(0) = \frac{1}{2}\ln\left[\frac{\bar{n}(0) + \frac{\bar{J}}{2v_s}\operatorname{sgn}(\bar{\psi}(0))}{\bar{p}(0) + \frac{\bar{J}}{2v_s}\operatorname{sgn}(\bar{\psi}(0))}\right].$$

Since we are only interested in a physically relevant continuous branch of solutions, which contains the reduced equilibrium solution $\bar{\psi} = \psi_{bi}$ for $\bar{J} = 0$, we choose $\operatorname{sgn}(\bar{\psi}(0)) = 1$. Therefore

$$\bar{\psi}(0) = \frac{1}{2}\ln\left[\frac{\bar{n}(0) + \frac{\bar{J}}{2v_s}}{\bar{p}(0) + \frac{\bar{J}}{2v_s}}\right] \tag{4.7.25}$$

holds, if

$$\bar{J} > -2v_s\bar{p}(0). \tag{4.7.26}$$

The layer problem, obtained by inserting (4.7.23) (b), (c) into (4.7.23) (a), reads:

$$\ddot{\hat{\psi}} = \left(\bar{n}(0) + \frac{\bar{J}}{2v_s}\right)e^{\hat{\psi}} - \left(\bar{p}(0) + \frac{\bar{J}}{2v_s}\right)e^{-\hat{\psi}} - 1, \qquad \omega > 0 \qquad \text{(4.7.27) (a)}$$

$$\hat{\psi}(0) = -\bar{\psi}(0), \qquad \hat{\psi}(\infty) = 0 \qquad \text{(4.7.27) (b)}$$

If (4.7.26) and $\bar{n}(0) > 0$ hold, then a straigthforward application of Lemma 4.5.1 implies the existence of a unique monotonically increasing solution $\hat{\psi}$. $|\hat{\psi}(\omega)|$ decays to zero exponentially as $\omega \to \infty$.
The reduced problem (4.7.20) can be solved by a lengthy but straightforward computation (see [4.22]). We obtain:

$$\bar{n}(x) = \frac{|\bar{J}| - v_s}{2v_s} + \sqrt{\left(\frac{v_s\sqrt{1+4\delta^4} - |\bar{J}|}{2v_s}\right)^2 + \frac{\bar{J}}{2}(1-x) + 1}, \qquad \text{(4.7.28) (a)}$$

$$\bar{p}(x) = \bar{n}(x) - 1 \qquad x \in [0, 1], \qquad \text{(4.7.28) (b)}$$

$$\bar{\psi}(x) = 2(\bar{n}(x) - \bar{n}(0)) + \bar{\psi}(0), \qquad x \in [0, 1], \qquad \text{(4.7.28) (c)}$$

where the reduced current density $\bar{J}$ is related to the applied bias U by the reduced voltage–current characteristic:

$$U = U(\bar{J}, v_s) \qquad \text{(4.7.28) (d)}$$

with

$$U(\bar{J}, v_s) = 2\ln\left[\frac{1+\sqrt{1+4\delta^4}}{2\delta^2}\right] - 2\sqrt{1+4\delta^4} + 2\frac{|\bar{J}|}{v_s}$$
$$+ 2\sqrt{\left(\frac{v_s\sqrt{1+4\delta^4} - |\bar{J}|}{v_s}\right)^2 + 2\bar{J}}$$
$$-\ln\left[\frac{1 + \frac{\bar{J}}{v_s}(\operatorname{sgn}(\bar{J})+1) + 2\sqrt{\left(\frac{v_s\sqrt{1+4\delta^4} - |\bar{J}|}{2v_s}\right)^2 + \frac{\bar{J}}{2}}}{-1 + \frac{\bar{J}}{v_s}(\operatorname{sgn}(\bar{J})+1) + 2\sqrt{\left(\frac{v_s\sqrt{1+4\delta^4} - |\bar{J}|}{2v_s}\right)^2 + \frac{\bar{J}}{2}}}\right]. \tag{4.7.29}$$

The expressions (4.7.28), however, only represent solutions, if the reduced current density $\bar{J}$ satisfies

$$J_{\min}(v_s) < \bar{J} < J_{\max}(v_s) \tag{4.7.30}$$

with:

$$J_{\max}(v_s) = v_s\sqrt{1+4\delta^4} \qquad \text{(4.7.31) (a)}$$

$$J_{\min}(v_s) = -\frac{4\delta^4 v_s}{v_s + \sqrt{1+4\delta^4} + \sqrt{v_s^2 + 2v_s\sqrt{1+4\delta^4}+1}}. \tag{4.7.31 (b)}$$

The lower bound stems from the inequality (4.7.26) and the upper bound is necessary for the solvability of the reduced problem (4.7.20).
A simple calculations gives

$$\lim_{\bar{J} \to J_{\min}(v_s)+} U(\bar{J}, v_s) = -\infty \tag{4.7.32 (a)}$$

and

$$\lim_{\bar{J} \to J_{\max}(v_s)-} U(\bar{J}, v_s) = 2\ln\left[\frac{1+\sqrt{1+4\delta^4}}{2\delta^2}\right] + 2\sqrt{2v_s\sqrt{1+4\delta^4}}$$

$$-\ln\left[\frac{1+2\sqrt{1+4\delta^4}+\sqrt{2v_s\sqrt{1+4\delta^4}}}{-1+2\sqrt{1+4\delta^4}+\sqrt{2v_s\sqrt{1+4\delta^4}}}\right] =: U_{\max}(v_s). \tag{4.7.32 (b)}$$

By differentiating $U(., v_s)$ with respect to J we find that $U(., v_s)$ is a strictly increasing and therefore invertible function of $\bar{J}$. Thus we have

$$\bar{J} = \bar{J}(U, v_s);$$

$$\bar{J}(., v_s)\colon (-\infty, U_{\max}(v_s)) \to (J_{\min}(v_s), J_{\max}(v_s)). \tag{4.7.33}$$

It can be shown that this branch of solutions terminates at $U_{\max}(v_s)$, i.e. it cannot be continuously extended beyond $U_{\max}(v_s)$. Note that

$$\lim_{\bar{J} \to J_{\max}-} \bar{n}'(1) = \lim_{\bar{J} \to J_{\max}-} \bar{p}'(1) = \lim_{\bar{J} \to J_{\max}-} \bar{\psi}'(1) = -\infty \tag{4.7.34}$$

holds.
The reduced solutions and their voltage-validity range depend decisively on the saturation velocity v_s. For fixed $|\psi'|$ we have

$$\lim_{v_s \to \infty} \frac{v_s}{v_s + |\psi'|} = 1,$$

i.e. the velocity-saturation device model (4.7.16) formally reduces to a constant-mobility model for infinite saturation velocity.
A simple calculation shows that

$$\lim_{v_s \to \infty} J_{\max}(v_s) = \lim_{v_s \to \infty} U_{\max}(v_s) = \infty$$

holds. The U-interval of existence $(-\infty, U_{\max}(v_s))$ of the branch of reduced solutions gets larger with increasing saturation velocity.
For the singularly perturbed problem (4.7.19) with the velocity-saturation mobility model (4.7.16) it is easy to show (by proceeding similarly to the proof of Theorem 3.5.1) that two continuous unbounded branches B^+, B^- of solutions, which both contain the equilibrium solution ($\psi = \psi_e$, $U = 0$) and at least one solution for every $U > 0$ and $U < 0$ resp., exist. The compactness argument employed for the proof of this result cannot be

carried over to the reduced problem due to the loss of the regularizing property of the inverse of the operator $L\psi = \psi''$. This facilitates the breakdown of the branch of reduced solutions at $U_{max}(v_s)$.
Numerical computations demonstrate that the 'full' voltage–current characteristic $J = J(U, v_s)$ saturates at a forward-saturation current density $J_s(v_s)$, which is of the order of magnitude of $J_{max}(v_s)$, i.e.

$$\lim_{U \to \infty} J(U, v_s) = J_s(v_s) \approx J_{max}(v_s)$$

(see [4.22]).
Since the residual of the zeroth order term of the asymptotic expansion (4.7.21) depends on the derivatives of the reduced solutions, which blow up at $x = 1$ as U tends to $U_{max}(v_s)$, we do not expect the sum of reduced and layer solutions to approximate the full solutions for U close to $U_{max}(v_s)$. We conjecture that the zeroth order terms of the expansion approximate the full solutions in forward bias if U is sufficiently smaller than $U_{max}(v_s)$.

4.8. Scaling Revisited and Conditioning

In this section we shall investigate the sensitivity of the device problem with respect to changes of the data. In the mathematical terminology an analysis of this kind is called a 'conditioning analysis'. A problem is well-conditioned, if small changes of the data cause small changes of the solutions and it is ill-conditioned, if small changes of the data can result in large changes of the solutions.
The conditioning of a 'continuous' problem determines to a large extent the conditioning of corresponding discretisations, i.e. reasonable discretisations of well-conditioned problems are well-conditioned and discretisations usually inherit ill-conditioning, too.
Ill-conditioning extremly complicates the computational solution of a problem. Small round-off errors are amplified and pollute or even destroy the simulation results. Also the speed of convergence of iterative methods for the approximate solution of nonlinear problems (like Newton's iteration or Gummel-type iterations, see Section 3.6) may be very slow such that the choice of initial guesses becomes extremly crucial.
We shall treat one-dimensional devices with Ohmic contacts here and only remark on the extension of the results to multi-dimensional cases.
At first we present an analysis of Poisson's equation, which highlights the effect of 'local' scaling (see Section 4.7) on the conditioning.

Rescaling Poisson's Equation

For given $u, v > 0$ the one-dimensional scaled Poisson's equation reads:

$$\lambda^2 \psi'' = \delta^2 e^{\psi} u - \delta^2 e^{-\psi} v - C(x), \qquad -1 \leqq x \leqq 1. \tag{4.8.1 (a)}$$

Dirichlet boundary conditions are prescribed at the contacts:

$$\psi(-1) = \psi_-, \qquad \psi(1) = \psi_+. \tag{4.8.1 (b)}$$

We pointed out in Section 4.7 that – for devices with strongly asymmetric junctions – 'local' scaling more clearly reveals the structure of solutions than the 'global' scaling which leads to (4.8.1). The following analysis demonstrates than an appropriate 'local' scaling of Poisson's equation is advantageous for numerical simulations.
We rescale the equation (4.8.1) (a) in such a way that the scaled doping profile is of the order of magnitude 1 everywhere with the possible exception of junction layers. This is achieved by premultiplying (4.8.1) (a) by an appropriate function, e.g. by $1/\sqrt{C^2(x)+4\delta^4}$ as suggested in [4.24a]. We define the rescaled quantities:

$$\lambda_s^2(x) := \frac{\lambda^2}{\sqrt{C^2(x)+4\delta^4}}, \qquad \delta_s^2(x) := \frac{\delta^2}{\sqrt{C^2(x)+4\delta^4}},$$
$$C_s(x) := \frac{C(x)}{\sqrt{C^2(x)+4\delta^4}}, \tag{4.8.2}$$

and obtain from (4.8.1) (a):

$$\lambda_s^2(x)\psi'' = \delta_s^2(x)e^{\psi}u - \delta_s^2(x)e^{-\psi}v - C_s(x), \qquad -1 \leqq x \leqq 1. \tag{4.8.3}$$

λ_s is a local normed Debye length (see Section 4.7) and δ_s a local scaled intrinsic number of the device.
We rewrite (4.8.3), (4.8.1) (b) as operator equation

$$P(\psi) = (C_s, \psi_+, \psi_-), \tag{4.8.4}$$

where the operator P is defined by:

$$P(\psi) = (\lambda_s^2\psi'' - (\delta_s^2 e^{\psi}u - \delta_s^2 e^{-\psi}v), \psi(-1), \psi(1)) \tag{4.8.5 (a)}$$

$$P\colon W^{2,\infty}(-1,1) \to L^\infty(-1,1)\times\mathbb{R}^2. \tag{4.8.5 (b)}$$

We equip $W^{2,\infty}(-1,1)$ with the weighted norm:

$$\|\psi\|_{\lambda_s} := \|\lambda_s^2\psi''\|_{\infty,(-1,1)} + \|\psi\|_{\infty,(-1,1)}. \tag{4.8.6}$$

By perturbing the (scaled) doping profile and the boundary data we obtain the equation:

$$P(\psi_1) = (C_{s,1}, \psi_{+,1}, \psi_{-,1}). \tag{4.8.7}$$

Taylor expansion gives:

$$P(\psi) - P(\psi_1) = D_\psi P(\psi)(\psi-\psi_1) - \frac{1}{2}D_\psi^2 P(\xi)(\psi-\psi_1)^2, \tag{4.8.8}$$

where ξ is on the line segment connecting ψ and ψ_1. Thus, neglecting second order terms, the error of the potential satisfies

$$D_\psi P(\psi)(\psi-\psi_1) \approx (C_s - C_{s,1}, \psi_- - \psi_{-,1}, \psi_+ - \psi_{+,1}). \tag{4.8.9}$$

and, consequently,

$$\|\psi-\psi_1\|_{\lambda_s} \lessapprox A_\psi(\|C_s-C_{s,1}\|_{\infty,(-1,1)}+|\psi_- - \psi_{-,1}|+|\psi_+ - \psi_{+,1}|), \tag{4.8.10}$$

where the amplification factor A_ψ is the norm of the inverse of the Frechet-derivative of P at the solution ψ:

$$A_\psi := \|(D_\psi P(\psi))^{-1}\|_{L^\infty(-1,1)\times\mathbb{R}^2\to W^{2,\infty}(-1,1)}. \tag{4.8.11}$$

We shall now derive a bound for A_ψ by estimating the solution φ of the linear equation

$$D_\psi P(\psi)\varphi = (f, \alpha, \beta), \tag{4.8.12}$$

which is represented by the two-point-boundary value problem:

$$\lambda_s^2\varphi'' = \frac{n+p}{\sqrt{C^2(x)+4\delta^4}}\varphi+f, \quad -1 \leqq x \leqq 1 \tag{4.8.13 (a)}$$

$$\varphi(-1) = \alpha, \quad \varphi(1) = \beta. \tag{4.8.13 (b)}$$

Typically, the carrier concentrations n, p and the doping profile C are – up to small errors – the sums of reduced terms $\bar{n}$, $\bar{p}$, $\bar{C}$ resp. and corresponding layer terms. For the sake of simplicity we shall neglect the influence of the layers and insert $\bar{n}$, $\bar{p}$, $\bar{C}$ for n, p, C resp. in (4.8.13) (a):

$$\lambda_s^2\sigma'' = \frac{\bar{n}+\bar{p}}{\sqrt{\bar{C}^2(x)+4\delta^4}}\sigma+f, \quad -1 \leqq x \leqq 1. \tag{4.8.14 (b)}$$

$$\sigma(-1) = \alpha, \quad \sigma(1) = \beta. \tag{4.8.14 (b)}$$

The maximum principle (see [4.23a]) implies the estimate:

$$\|\sigma\|_{\infty,(-1,1)} \leqq |\alpha|+|\beta|+\sup_{(-1,1)}\frac{\sqrt{\bar{C}^2(x)+4\delta^4}}{\bar{n}+\bar{p}}\|f\|_{\infty,(-1,1)}. \tag{4.8.15}$$

From (4.4.4) we derive $\bar{n}+\bar{p} = \sqrt{\bar{C}^2+4\delta^4\bar{u}\bar{v}}$ and assuming that $\inf_{(-1,1)} |\bar{C}(x)| \geqq \delta^2$, we obtain

$$\frac{\sqrt{C^2+4\delta^4}}{\bar{n}+\bar{p}} \leqq 5. \tag{4.8.16}$$

By a more sophisticated application of the maximum principle it can be shown – at least for abrupt doping profiles – that internal layers give rise to an additional factor of the order of magnitude $|\ln\delta|$ in the estimate (4.8.15) (see [4.17]). Thus, we obtain the following estimate for the amplification factor A_ψ:

$$A_\psi \leqq \text{const.}\ (1+|\ln\delta|)\left(1+\sup_{(-1,1)}\frac{\bar{n}+\bar{p}}{\sqrt{\bar{C}^2+4\delta^4}}\right), \tag{4.8.17}$$

where the constant is independent of λ, δ and C. We shall estimate the last term on the right hand side of (4.8.17) below.
The so called condition number

$$C_\psi = A_\psi B_\psi, \tag{4.8.18}$$

where B_ψ is defined by

$$B_\psi := \|D_\psi P(\psi)\|_{W^{2,\infty}(-1,1)\to L^\infty(-1,1)\times\mathbb{R}^2}, \tag{4.8.19}$$

provides an appropriate measure of the conditioning of the linearized Poisson's equation (4.8.12) (see [4.8a]). Using the definition of the norm (4.8.6) and, again, ignoring the influence of layers we obtain, evoking (4.4.2) and (4.4.4):

$$B_\psi \lessapprox \sup_{(-1,1)} \frac{\bar{n}(x)+\bar{p}(x)}{\sqrt{\bar{C}^2(x)+4\delta^4}}+1$$

$$\leqq \sup_{(-1,1)} \sqrt{\frac{\bar{C}^2(x)+4\delta^4 \exp(|U|)}{\bar{C}^2(x)+4\delta^4}}+1 \leqq \exp\left(\frac{|U|}{2}\right)+1 \tag{4.8.20}$$

assuming that the recombination-generation rate is of SHR- and/or Auger type. U denotes the externally applied bias. Thus, the estimate of the condition number reads:

$$C_\psi \lessapprox \text{const.} \exp(|U|)(1+|\ln\delta|). \tag{4.8.21}$$

Poisson's equation (4.8.3) is – at least for moderate applied bias – well conditioned independently of the singular perturbation parameter λ and of the doping profile. The condition number only depends weakly on δ.
Clearly, the 'local' scaling has no impact on the exact solution of Poisson's equation, it only provides a sensible way of measuring perturbations of the right hand side. Local absolute errors of C_s correspond to local relative errors of the unscaled doping profile (away from device junctions). Apparently, the optimal local scaling of Poisson's equation is obtained by dividing (4.8.1) (a) by $n+p$ instead of $\sqrt{C^2(x)+4\delta^4}$, since in this case the condition number of the linearisation equals 1. Then local absolute errors of the scaled doping profile correspond to local errors of the unscaled doping profile relative to the local total unscaled carrier concentration.
Appropriately measuring perturbations of the right hand side is particularly important when termination criteria for nonlinear iterations are based on the magnitude of residuals (see [4.24a] for an application in context with Newton's method). A practical 'close to optimal' scaling is realised by dividing (4.8.1) (a) by n_l+p_l, where n_l, p_l are the iterates obtained in the last Newton- or Gummel-step (see Section 3.6).

Conditioning of the Continuity Equations

We set $\mu_n \equiv 1$ and assume that the recombination-generation rate R and the potential ψ are given. Then – in the one-dimensional case – the electron concentration n solves the two-point boundary value problem:

$$(n' - n\psi') = R(x), \qquad -1 \leqq x \leqq 1 \tag{4.8.22 (a)}$$

$$n(-1) = n_-, \qquad n(1) = n_+. \tag{4.8.22 (b)}$$

We regard the boundary values n_+, n_- and the recombination-generation rate R as data. Since the problem (4.8.22) is linear in n the sensitivity of the solution with respect to changes of the data can be expressed by estimating n in terms of n_-, n_+ and R.
We prove:

Theorem 4.8.1 *The solution n of (4.8.22) satisfies the estimate*

$$\|n\|_{\infty,(-1,1)} \leqq e^{\psi_{\sup} - \psi_{\inf}}(|n_-| + |n_+| + 4\|R\|_{1,(-1,1)}), \tag{4.8.23}$$

where $\psi_{\sup} := \sup_{(-1,1)} \psi(x)$, $\psi_{\inf} := \inf_{(-1,1)} \psi(x)$.

Proof: The problem (4.8.22) can be solved explicitely by straightforward integration:

$$n(x) = \frac{\int_x^1 e^{\psi(x)-\psi(s)}\,ds}{\int_{-1}^1 e^{\psi(-1)-\psi(s)}\,ds}\, n_- + \frac{\int_{-1}^x e^{\psi(x)-\psi(s)}\,ds}{\int_{-1}^1 e^{\psi(1)-\psi(s)}\,ds}\, n_+ + n_{pa}[R]\,(x) \tag{4.8.24 (a)}$$

$$n_{pa}[R]\,(x) = \int_{-1}^x e^{\psi(x)-\psi(s)} \int_{-1}^s R(\tau)\,d\tau\,ds - \int_{-1}^1 e^{\psi(x)-\psi(s)} \int_{-1}^s R(\tau)\,d\tau\,ds\, \frac{\int_{-1}^x e^{-\psi(s)}\,ds}{\int_{-1}^1 e^{-\psi(s)}\,ds}. \tag{4.8.24 (b)}$$

The assertion of the theorem is obtained by estimating (4.8.24) (a), (b). □

An estimate for the current density J_n can easily be derived, too.

Corollary 4.8.1 *The electron current density* $J_n = n' - n\psi'$ *satisfies:*

$$\|J_n\|_{\infty,(-1,1)} \leqq \frac{e^{-\psi(-1)}}{\int_{-1}^1 e^{-\psi(s)}\,ds}|n_-| + \frac{e^{-\psi(1)}}{\int_{-1}^1 e^{-\psi(s)}\,ds}|n_+| + 2\|R\|_{1,(-1,1)}. \tag{4.8.25}$$

Proof: We obtain from (4.8.24):

$$J_n(x) = \frac{n_+ e^{-\psi(1)} - n_- e^{-\psi(-1)}}{\int_{-1}^{1} e^{-\psi(s)}\,ds} + \int_{-1}^{x} R(s)\,ds - \frac{\int_{-1}^{1} e^{-\psi(s)} \int_{-1}^{s} R(\tau)\,d\tau\,ds}{\int_{-1}^{1} e^{-\psi(s)}\,ds}. \tag{4.8.26}$$

The estimate (4.8.25) follows immediately. □

Theorem 4.8.1 implies that perturbations of the data of the continuity equation (4.8.22) are amplified maximally by the factor $4e^{\psi_{\sup} - \psi_{\inf}}$ in the electron concentration n. By using the bounds of the potential given in Section 3.2 we obtain the following estimate for the amplification factor:

$$A_n := 4e^{\psi_{\sup} - \psi_{\inf}} \leqq 4\,\frac{|C_{\sup}|\,|C_{\inf}|}{\delta^4}\,e^{|U|}\left(1 + O\left(\left(\frac{\delta^2}{|C_{\sup}|}\right)^2\right) + O\left(\left(\frac{\delta^2}{|C_{\inf}|}\right)^2\right)\right), \tag{4.8.27}$$

where we set $C_{\sup} := \sup\limits_{(-1,1)} C(x)$, $C_{\inf} := \inf\limits_{(-1,1)} C(x)$. Thus, we have

$$A_n = O\left(\frac{e^{|U|}}{\delta^4}\right). \tag{4.8.28}$$

For a silicon device with a maximal doping concentration $10^{17}\,\mathrm{cm}^{-3}$ we calculate $\delta^2 \approx 10^{-7}$ and, therefore, the upper bound for A_n is – assuming even that the device is in thermal equilibrium – of the order of magnitude 10^{14}.

We shall now investigate whether the estimate (4.8.23) is sharp, i.e. whether there are (practically relevant) devices for which the amplification factor is actually of the order of magnitude $1/\delta^4$.

First, we consider a *pnp*-device with a piecewise homogeneous doping profile

$$C(x) = \begin{cases} C_-, & -1 \leqq x < -\frac{1}{2} \\ C_+, & -\frac{1}{2} < x < \frac{1}{2}, \\ C_-, & \frac{1}{2} < x \leqq 1 \end{cases} \qquad C_- < 0, \quad C_+ > 0. \tag{4.8.29}$$

In thermal equilibrium the potential ψ is well approximated by the reduced potential $\bar{\psi}$ except in thin layer strips about the junctions $x = \pm 1/2$. We easily calculate:

$$\bar{\psi}(x) = \begin{cases} \psi_-, & -1 \leqq x < -\dfrac{1}{2} \\ \psi_+, & -\dfrac{1}{2} < x < \dfrac{1}{2}, \\ \psi_-, & \dfrac{1}{2} < x \leqq 1 \end{cases} \qquad \psi_\pm := \ln\left[\frac{C_\pm + \sqrt{C_\pm^2 + 4\delta^4}}{2\delta^2}\right], \tag{4.8.30}$$

Clearly, $\psi_- < 0$ and $\psi_+ > 0$.
For the sake of simplicity we ignore the influence of the layers and insert $\bar{\psi}$ for ψ into (4.8.24). Formally, this corresponds to solving the reduced continuity equation. We obtain the reduced electron concentration:

$$\bar{n}(x) = \begin{Bmatrix} \dfrac{e^{\psi_- - \psi_+} - x}{e^{\psi_- - \psi_+} + 1} n_- + \dfrac{x+1}{e^{\psi_- - \psi_+} + 1} n_+, & -1 \leqq x < -\dfrac{1}{2} \\ \dfrac{\frac{1}{2} - x + \frac{1}{2} e^{\psi_+ - \psi_-}}{e^{\psi_- - \psi_+} + 1} n_- + \dfrac{x + \frac{1}{2} + \frac{1}{2} e^{\psi_+ - \psi_-}}{e^{\psi_- - \psi_+} + 1} n_+, & -\dfrac{1}{2} \leqq x < \dfrac{1}{2} \\ \dfrac{1-x}{e^{\psi_- - \psi_+} + 1} n_- + \dfrac{e^{\psi_- - \psi_+} + x}{e^{\psi_- - \psi_+} + 1} n_+, & \dfrac{1}{2} < x \leqq 1 \end{Bmatrix} + \bar{n}_{\mathrm{pa}}[R](x) \tag{4.8.31}$$

and evaluate the particular solution at $x = 0$:

$$\bar{n}_{\mathrm{pa}}[R](0) = \frac{1}{2}\left(e^{\psi_+ - \psi_-}\left(\int_{-1}^{-\frac{1}{2}} \int_{-1}^{s} R(\tau)\,\mathrm{d}\tau\, ds - \int_{\frac{1}{2}}^{1} \int_{-1}^{s} R(\tau)\,\mathrm{d}\tau\, ds\right) + \right.$$
$$\left. + \int_{-\frac{1}{2}}^{0} \int_{-1}^{s} R(\tau)\,\mathrm{d}\tau\, ds - \int_{0}^{\frac{1}{2}} \int_{-1}^{s} R(\tau)\, d\tau\, ds\right). \tag{4.8.32}$$

The coefficient functions of n_+, n_- are of the order of magnitude $e^{\psi_+ - \psi_-}$ in the n-domain $(-1/2, 1/2)$, they are $0(1)$ in the two p-domains.

Also, the particular solution is $0(e^{\psi_+ - \psi_-})$ at $x = 0$. Thus, by evoking (4.8.30), we conclude that the estimate (4.8.28) is sharp for this particular device in thermal equilibrium (when layer effects are neglected). The amplification factor is of the order of magnitude $1/\delta^4$.
As second device example we take a pn-diode:

$$C(x) = \begin{cases} C_-, & -1 \leqq x < 0 \\ C_+, & 0 < x \leqq 1 \end{cases}, \qquad C_- < 0, \qquad C_+ > 0. \tag{4.8.33}$$

The reduced potential in thermal equilibrium is given by:

$$\bar{\psi}(x) = \begin{cases} \psi_- < 0, & -1 \leqq x < 0 \\ \psi_+ > 0, & 0 < x \leqq 1 \end{cases}. \tag{4.8.34}$$

By inserting into (4.8.24) we obtain the reduced electron concentration

$$\bar{n}(x) = \begin{cases} \dfrac{e^{\psi_- - \psi_+} - x}{e^{\psi_- - \psi_+} + 1} n_- + \dfrac{x+1}{e^{\psi_+ - \psi_-} + 1} n_+ + \int\limits_{-1}^{x} \int\limits_{-1}^{s} R(\tau)\, d\tau\, ds \\ - \left(\int\limits_{-1}^{0} \int\limits_{-1}^{s} R(\tau)\, d\tau\, ds \right. \\ \left. + e^{\psi_- - \psi_+} \int\limits_{0}^{1} \int\limits_{-1}^{s} R(\tau)\, d\tau\, ds \right) \dfrac{x+1}{e^{\psi_- - \psi_+} + 1}, \quad -1 \leqq x < 0 \\ \dfrac{1-x}{e^{\psi_- - \psi_+} + 1} n_- + \dfrac{x + e^{\psi_+ - \psi_-}}{e^{\psi_+ - \psi_-} + 1} n_+ + \int\limits_{0}^{x} \int\limits_{0}^{s} R(\tau)\, d\tau\, ds + \\ + \dfrac{1-x}{e^{\psi_- - \psi_+} + 1} \int\limits_{-1}^{0} \int\limits_{-1}^{s} R(\tau)\, d\tau\, ds \\ + \dfrac{1 + x e^{\psi_- - \psi_+}}{e^{\psi_- - \psi_+} + 1} \int\limits_{0}^{1} \int\limits_{-1}^{s} R(\tau)\, d\tau\, ds, \quad 0 < x \leqq 1, \end{cases} \tag{4.8.35}$$

and the estimate

$$\|\bar{n}\|_{\infty,(-1,1)} \leqq |n_-| + |n_+| + 3\, \|R\|_{1,(-1,1)} \tag{4.8.36}$$

follows. The amplification factor of the reduced n-continuity equation for the *pn*-diode in thermal equilibrium is maximally 3, i.e. it is bounded independently of δ.

A straightforward calculation shows that the amplification factors of the corresponding reduced p-continuity equations for the *pnp*-device *and* for the *pn*-diode are bounded independently of δ. For one-dimensional devices with more than two *pn*-junctions, however, the amplification factors of both reduced continuity equations are of the order of magnitude $1/\delta^4$ in thermal equilibrium.

In the one-dimensional case the ill-conditioning is caused by the occurence of n- or p-regions without contacts. By heuristically extrapolating this result we conjecture that – sufficiently close to thermal equilibrium – two cases have to be distinguished.

(a) Both continuity equations are well-conditioned independently of δ, if every p- and n-region, which occurs in the device under consideration, has a contact.

(b) If an n- or p-region has no contact, then the continuity equation for the majority carrier concentration of this region is ill-conditioned. Its amplification factor is of the order of magnitude $1/\delta^4$.

The two-dimensional *pn*-diode of Fig. 4.6.1 represents an example for the case (a). Both continuity equations are well-conditioned sufficiently close to thermal equilibrium since the n- and the p-region are contacted. The situation

is different in the case of the thyristor depicted in Fig. 4.3.1. The middle n-region has no contact, thus – close to thermal equilibrium – we expect the n-continuity equation to be ill-conditioned and the p-continuity to be well conditioned (since all p-regions are contacted).
The upper bound for the amplification factor A_n given in (4.8.28) increases as the absolute value of the applied bias increases. However, this does not necessarily reflect accurately the impact of the applied bias on the conditioning of the continuity equations. If at least one device junction is strongly reverse biased, then $\psi_{\text{sup}} - \psi_{\text{inf}}$ is larger than $(\psi_e)_{\text{sup}} - (\psi_e)_{\text{inf}}$ and the conditioning is likely to be worse than in thermal equilibrium. If all junctions are forward biased, then $\psi_{\text{sup}} - \psi_{\text{inf}}$ is usually smaller than $(\psi_e)_{\text{sup}} - (\psi_e)_{\text{inf}}$ and the conditioning may be even better than in equilibrium.
For devices, which lack a sufficient stabilisation effect of contacts, perturbations in the data of the continuity equations may appear amplified by (the large) factor $1/\delta^4$ in the carrier concentrations. Normally, the effect of perturbations on the current densities is less dramatic. Corollary 4.8.1 implies that $L^1(-1, 1)$-perturbations of the recombination-generation rate are at most amplified by the factor 2. Also, the coefficient functions of the boundary data are $O(1)$ unless the potential decays very steeply away from the contacts. At least in moderate injection small perturbations of the data cause small perturbations of the current densities even if the perturbations of the carrier concentrations are large.
If the recombination-generation rate is of the form $R = Q(x, n, p)(np - \delta^4)$, $Q \geqq 0$, $Q \not\equiv 0$ then – more realistically – the conditioning of the modified linear operator

$$\begin{aligned} &L(\psi_0, n_0, p_0)n_1 \\ &:= ((n_1' - n_1\psi_0')' - Q(x, n_0, p_0)p_0 n_1, n_1(-1), n_1(1)) \end{aligned} \tag{4.8.37}$$

should be investigated. Clearly, the term Qp_0 has a stabilising effect, however, for the SRH and for the Auger recombination-generation terms we have

$$\inf_{\Omega} (Q(x, n, p)p) = O(\delta^4) \tag{4.8.38}$$

in low injection, and thus, a significant improvement of the conditioning cannot be expected close to thermal equilibrium.
Since the conditioning is inherited by reasonable discretisations, there are important practical consequences of the analysis presented above.
When Gummel-type iterations are employed (see Section 3.6) then linear problems of the type (4.8.13) and (4.8.22) or, resp., (4.8.37) have to be solved in every iteration step. The linearized Poisson's equation (4.8.22) is well conditioned and – depending on the device under consideration – the continuity equations may be ill- or well-conditioned. In the ill-conditioned case various difficulties in the numerical solution are encountered. First of all, the convergence of the Gummel-type iteration is usually very slow such that the choice of a good initial guess, obtained by, e.g., a continuation method (see [4.16a]), is of paramount importance. Also, in each Gummel step, the numerical algorithm for solving the discretised continuity equations is strongly influenced by their ill-conditioning. Iterative methods, e.g. relaxation meth-

ods (see [4.8a]), for the approximate solution of the linear systems may converge very poorly such that direct methods (Gaussian elimination) are preferable.
The ill-conditioning of the continuity equations has an analogous impact on Newton-type iterations.

References

[4.1] Ascher, U., Christiansen, J., Russell, R. D.: A Collocation Solver for Mixed Order Systems of Boundary Value Problems. Math. Comp. *33,* 659–679 (1979).

[4.2] Ascher, U.: Solving Boundary Value Problems with a Spline Collocation Code. J. Comp. Phys. *34,* 401–413 (1980).

[4.3] Ashcroft, N. W., Mermin, N. D.: Solid State Physics. Philadelphia: Saunders 1976.

[4.3a] Brezzi, F., Capelo, A., Marini, L. D.: Singular Perturbation Problems in Semiconductor Devices. Report 464, Istituto di Analisi Numerica del Consiglio Nazionale delle Ricerche, Pavia, Italy, 1985.

[4.3b] Caffarelli, L. A., Friedman, A.: A Singular Perturbation Problem for Semiconductors. (To appear, 1986.)

[4.4] Cole, J. D.: Perturbation Methods in Applied Mathematics. Blaisdell 1968.

[4.5] De Hoog, F. R., Weiss, R.: An Approximation Method for Boundary Value Problems on Infinite Intervals. Computing *24,* 227–239 (1980).

[4.6] Eckhaus, W.: Asymptotic Analysis of Singular Perturbations. Amsterdam–New York–Oxford: North-Holland 1979.

[4.7] Fife, P. C.: Semilinear Elliptic Boundary Value Problems with Small Parameters. Arch. Rational Mech. *52,* 205–232 (1974).

[4.8] Franz, A. F., Franz, G. A., Selberherr, S., Ringhofer, C., Markowich, P. A.: Finite Boxes: A Generalisation of the Finite Difference Method Suitable for Semiconductor Device Simulation. IEEE Transactions on Electron Devices. *ED–30,* No. 9, 1070–1083 (1983).

[4.8a] Golub, G. H., Van Loan, C. F.: Matrix Computations. Baltimore, Johns Hopkins University Press 1983.

[4.9] Howes, F. A.: Boundary-Interior Layer Interactions in Nonlinear Singular Perturbation Theory. Memoirs of the AMS *15,* No. 203 (1978).

[4.10] Howes, F. A.: Robin and Neumann Problems for a Class of Singularly Perturbed Semilinear Elliptic Equations. J. of Differential Equations *34,* 55–73 1979.

[4.11] Ladyshenskaja, O. A., Uraltseva, N. N.: Linear and Quasilinear Elliptic Equations. New York: Academic Press 1968.

[4.12] Lin, C. C., Segel, L. A.: Mathematics Applied to Deterministic Problems in the Natural Sciences. New York: Macmillan 1974.

[4.13] O'Malley, R. E., jr.: Introduction to Singular Perturbations. New York: Academic Press 1974.

[4.14] Markowich, P. A.: Analysis of Boundary Value Problems on Infinite Intervals. SIAM J. Math. Anal. *14,* 11–37 (1983).

[4.15] Markowich, P. A.: A Theory for the Approximation of Solutions of Boundary Value Problems on Infinite Intervals. SIAM J. Math. Anal. *13,* No. 3, 484–513 (1982).

[4.16] Markowich, P. A., Ringhofer, C. A.: A Singularly Perturbed Boundary Value Problem Modelling a Semiconductor Device. SIAM J. Appl. Math. *44,* No. 2, 231–256 (1984).

[4.16a] Markowich, P. A., Ringhofer, C. A., Steindl, A.: Computation of Current-Voltage Characteristics in a Semiconductor Device using Arc-Length Continuation. IMA J. Appl. Math. *33,* 175–187 (1984).

[4.17] Markowich, P. A., Ringhofer, C. A., Selberherr, S., Langer, E.: An Asymptotic Analysis of Single Junction Semiconductor Devices. MRC–TSR 2527, Math. Res. Center, University of Wisconsin–Madison, U.S.A., 1983.

[4.18] Markowich, P. A.: A Singular Perturbation Analysis of the Fundamental Semiconductor Device Equations. SIAM J. Appl. Math. *5,* No. 44, 896–928 (1984).

[4.19] Markowich, P. A.: A Qualitative Analysis of the Fundamental Semiconductor Device Equations. COMPEL *2,* No. 3, 97–115 (1983).

[4.20] Markowich, P. A., Ringhofer, C. A., Selberherr, S., Lentini, M.: A Singular Perturbation Approach for the Analysis of the Fundamental Semiconductor Equations. IEEE Transactions on Electron Devices. *ED–30*, No. 9, 1165–1181 (1983).

[4.21] Markowich, P. A., Schmeiser, C.: Asymptotic Representation of Solutions of the Basic Semiconductor Device Equations. MRC–TSR 2772, Math. Res. Center, University of Wisconsin–Madison, U.S.A., 1985.

[4.22] Markowich, P. A., Schmeiser, C.: A Singular Perturbation Analysis of One-Dimensional Semiconductor Device Models with Velocity Saturation Effects. Report, Institut für Angewandte und Numerische Mathematik, Technische Universität Wien, Austria, 1985.

[4.23] Please, C. P.: An Analysis of Semiconductor P–N Junctions. IMA J. of Appl. Math. *28*, 301–318 (1982).

[4.23a] Protter, M. H., Weinberger, H. F.: Maximum Principles in Differential Equations. Englewood Cliffs, N. J.: Prentice-Hall 1967.

[4.24] Schmeiser, C., Weiss, R.: Asymptotic Analysis of Singular Singularly Perturbed Boundary Value Problems. SIAM J. Math. Anal. (1985).

[4.24a] Schmeiser, C., Selberherr, S., Weiss, R.: On Scaling and Norms for Semiconductor Device Simulation. In: Proc. NASECODE IV Conference. Dublin: Boole Press 1985.

[4.25] Selberherr, S.: Two Dimensional Modeling of MOS-Transistors. Published by Semiconductor Physics, Inc., 639 Meadow Grove Place, Escondido, Cal., U.S.A., 1982.

[4.26] Selberherr, S., Ringhofer, C. A.: Implications of Analytical Investigations about the Semiconductor Equations on Device Modeling Programs. IEEE Transactions on Computer Aided Design. *CAD–3*, No. 1, 52–64 (1984).

[4.27] Selberherr, S.: Analysis and Simulation of Semiconductor Devices. Wien–New York: Springer 1984.

[4.28] Smith, D.: On a Singularly Perturbed Boundary Value Problem Arising in the Physical Theory of Semiconductors. Report, Institut für Mathematik und Informatik, Technische Universität München, München, FRG, 1980.

[4.29] Sze, S. M.: Physics of Semiconductor Devices, 2nd ed. Cambridge–New York: Wiley-Interscience 1981.

[4.30] Vasileva, A. B., Butuzov, V. F.: Singularly Perturbed Equations in the Critical Case. MRC–TSR 2039, Math. Res. Center, University of Wisconsin–Madison, U.S.A., 1980.

[4.31] Vasileva, A. B., Stelmakh, V. G.: Singularly Disturbed Systems of the Theory of Semiconductor Devices. USSR Comput. Math. Phys. *17*, 48–58 (1977).

Discretisation of the Stationary Device Problem 5

The numerical solution of boundary value problems for nonlinear systems of elliptic partial differential equations in general and the static simulation of semiconductor devices in particular usually proceeds in the following steps:

(i) The 'continuous' problem is replaced by a suitable approximating 'discrete' nonlinear system of algebraic equations, whose solutions are intrinsically related to point-values of approximate solutions. This procedure is called discretisation of the boundary value problem.
(ii) Since, usually, the nonlinear system of equations generated by the discretisation cannot be solved exactly, an iteration scheme based on (quasi-) linearisation is set up in order to obtain an approximate discrete solution.
(iii) In each iteration step a usually large and sparse system of linear equations has to be solved.

In this chapter we shall derive and analyse discretisations of the stationary device problem. We shall also remark on the steps (i) and (ii), however, an in-depth analysis and assessment of solution strategies for systems of algebraic equations is beyond the scope of this book. We refer the interested reader to the references [5.26] and [5.32] for excellent mathematical treatments and to [5.29] for a survey of methods and an assessment of their applicability and performance in context with the semiconductor device problem.
The two most important philosophies for the construction of discretisation schemes are represented by the finite element and by the finite difference approach. A finite element discretisation is based on approximating the weak (variational) formulation of the boundary value problem in a finite dimensional subspace of the basic infinite dimensional function space (Ritz–Galerkin method). Usually the finite dimensional subspace consists of piecewise polynomials on a partition of the domain Ω. The (small) subregions, which

make up the partion, are called finite elements and the partition represents the finite element mesh. Normally, a basis of the finite dimensional subspace consisting of functions, which have a 'small' support (a few finite elements), can be determined and the Ritz–Galerkin approximation gives a sparse finite dimensional system of equations.

A (classical) finite difference discretisation of a boundary value problem is obtained by covering the domain Ω by a usually fine rectangular grid and by replacing all derivatives in the strong formulation of the differential equation by difference quotients at the gridpoints. In this way a sparse finite dimensional approximating system of equations is obtained, too.

Although these two approaches are totally different in their origins, it is well known that the finite difference schemes commonly used for discretising elliptic boundary value problems are equivalent (up to small perturbations) to finite element discretisations, when the finite dimensional subspace and the finite element mesh are chosen appropriately (see [5.19]). Therefore we shall derive suitable finite element methods for the static device problem and obtain finite difference schemes as special cases.

Difference methods are often used in semiconductor device simulation packages because they require relatively reasonable coding and manpower resources, however finite element methods allow greater flexibility in adjusting the mesh to the structure of solutions.

A main issue of the subsequent analysis is the determination of the accuracy of the discretisation schemes in dependence on the finite element mesh or, resp., on the finite difference grid. The accuracy analysis in conjunction with the singular perturbation analysis of Chapter 4, which revealed the layer structure of solutions, will lead us to the qualitative construction of a-priori meshes admitting reasonable predefined error distributions. We shall devise an efficient griding strategy for finite difference schemes, which allows local grid refinement within layer regions and therefore partly overcomes the disadvantage of finite difference methods (in comparison to finite element methods).

We refer those readers, who are interested in the basics and in the analysis of finite element methods, to the books [5.1], [5.6], [5.25], [5.30] and [5.34]. Introductions to the derivation and analysis of finite difference schemes can be found in [5.10], [5.19]. A phenomenological analysis of discretisations of the static device problem is given in [5.29] and a mathematical analysis, focusing on classical convergence estimates, in [5.21].

5.1 Construction of Discretisations

In this section we shall construct finite element discretisations of the static device problem and derive finite difference schemes as special cases.

For reasons of simplicity we assume that the boundary $\partial\Omega$ of the k-dimensional domain Ω ($k = 1$, 2 or 3) splits into Neumann segments $\partial\Omega_N$ and Dirichlet segments $\partial\Omega_D$ only. $\partial\Omega_N$ is open in $\partial\Omega$ and $\partial\Omega_D$ is closed in $\partial\Omega$. The $(k-1)$-dimensional Lebesguemeasure of $\partial\Omega_D$ is positive. In Section 5.4 we shall admit semiconductor-oxide interfaces and Schottky contacts.

Discretisation of Poisson's Equation

A. The Ritz–Galerkin Approximation

For given carrier concentrations n and p the potential ψ solves Poisson's equation:

$$\lambda^2 \Delta\psi = n - p - C(x), \qquad x \in \Omega, \tag{5.1.1 (a)}$$

subject to the boundary conditions:

$$\left.\frac{\partial\psi}{\partial\nu}\right|_{\partial\Omega_N} = 0, \qquad \psi|_{\partial\Omega_D} = \psi_D|_{\partial\Omega_D}. \tag{5.1.1 (b)}$$

We assume that the Dirichlet datum $\psi_D \in H^1(\Omega)$ and the doping profile $C \in L^\infty(\Omega)$.
The weak formulation of (5.1.1) is obtained by multiplying (5.1.1) (a) with an arbitrary test function $\varphi \in H_0^1(\Omega \cup \partial\Omega_N)$ and integrating by parts:

$$-\lambda^2 a(\psi, \varphi) = (n - p - C, \varphi)_{2,\Omega}$$

$$\text{for all } \varphi \in H_0^1(\Omega \cup \partial\Omega_N), \tag{5.1.2 (a)}$$

$$\psi - \psi_D \in H_0^1(\Omega \cup \partial\Omega_N), \tag{5.1.2 (b)}$$

where $a(.,.)$ denotes the coercive, bounded and symmetric bilinear form on $H^1(\Omega)$:

$$a(\psi, \varphi) = \int_\Omega \operatorname{grad}\psi \cdot \operatorname{grad}\varphi \, dx; \qquad \varphi, \psi \in H^1(\Omega) \tag{5.1.3}$$

and $(.,.)_{2,\Omega}$ is the $L^2(\Omega)$-inner product.
Let X denote a finite dimensional subspace of $H_0^1(\Omega \cup \partial\Omega_N)$. Then the Ritz–Galerkin approximation of (5.1.2) (a), (b) on X is defined by (see, e.g., [5.1]):

$$-\lambda^2 a(\psi_a, \varphi) = (n - p - C, \varphi)_{2,\Omega} \qquad \text{for all } \varphi \in X, \tag{5.1.4 (a)}$$

$$\psi_a - \psi_D \in X. \tag{5.1.4 (b)}$$

For given $n, p \in L^2(\Omega)$ the existence of a unique Ritz–Galerkin solution $\psi_a \in X$ follows immediately by applying the Lax–Milgram Lemma (see, e.g., [5.12]).

B. Piecewise Linear Finite Element Approximation

A way of constructing suitable finite dimensional subspaces X is provided by the finite element method. Therefore we partition the region $\bar{\Omega}$ into subregions (so called finite elements) and define the functions of the subspace X locally on these subregions.
In the one-dimensional case $\Omega = (-1, 1)$ the subregions are subintervals:

$$\bar{\Omega} = \bigcup_{l=1}^{N} T_l, \qquad T_l = [x_{l-1}, x_l], \tag{5.1.5 (a)}$$

$$-1 = x_0 < x_1 < \dots < x_{N-1} < x_N = 1. \tag{5.1.5 (b)}$$

In the two and three-dimensional cases various ways of partitioning $\bar{\Omega}$ lead to reasonable finite element approximations, very often triangles in two dimensions and tetrahedrons in three dimensions are used.
For the following we assume that the boundary $\partial\Omega$ consists of a union of line ($k = 2$) or, resp., plane segments ($k = 3$). In the three-dimensional case $\partial\Omega_N$ and $\partial\Omega_D$ are also assumed to consist of a union of plane segments. These assumptions hold for the commonly used device simulation domains, the boundaries and contact edges of more complicated domains have to be approximated by piecewise linear surfaces.
We partition $\bar{\Omega}$ into N closed triangles (see Fig. 5.1.1) or resp., N closed tetrahedrons, which we denote by T_l, $l = 1, \dots, N$:

$$\bar{\Omega} = \bigcup_{l=1}^{N} T_l, \qquad \Delta := \{T_1, \dots, T_N\}. \tag{5.1.6}$$

The partition Δ is chosen such that two different elements either have an empty intersection, or they have a vertex, a whole edge or – in three dimensions – a whole side surface in common. Also those elements, which have a nonempty intersection with $\partial\Omega_D$ are assumed to have an empty intersection with $\partial\Omega_N$.

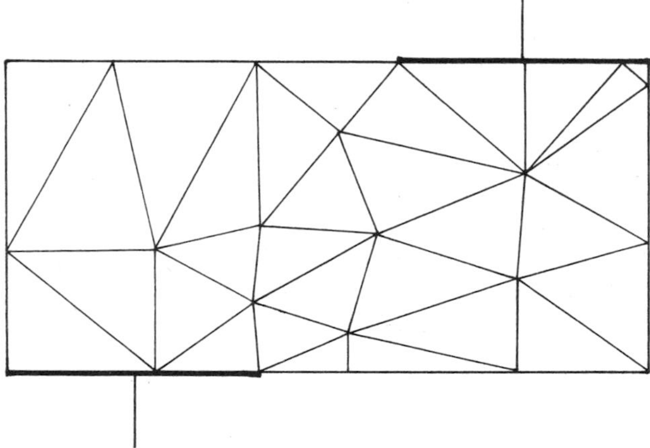

Fig. 5.1.1 Triangulation of $\bar{\Omega}$

We shall use the following notations:

N: number of elements in the partition
K: number of vertices in $\bar{\Omega}$
M: number of vertices in $\Omega \cup \partial\Omega_N$
L: number of vertices in Ω
$\{P_1, \dots, P_L\} \subseteq \Omega$: set of interior vertices
$\{P_{L+1}, \dots, P_M\} \subseteq \Omega_N$: set of Neumann boundary vertices
$\{P_{M+1}, \dots, P_K\} \subseteq \Omega_D$: set of Dirichlet boundary vertices
$d_l := \operatorname{diam}(T_l)$: diameter of the element T_l
$d := \max_{l=1,\dots,N} d_l$: diameter of the partition Δ.

We define the subspace X_Δ of $H^1(\Omega)$ as the space of all functions, which are continuous on $\bar{\Omega}$ and which are linear on each of the finite elements T_l. The subspace $X_{\Delta,0}$ of $H_0^1(\Omega \cup \partial\Omega_N)$ is defined as the space of all functions, which are in X_Δ and which vanish on (the closed boundary subset) $\partial\Omega_D$. (Note that the assumptions on the partition Δ imply that the definition of $X_{\Delta,0}$ as subset of X_Δ makes sense). The dimensions of X_Δ and $X_{\Delta,0}$ are K and M resp. A basis of X_Δ is given by:

$$B_\Delta := \{\varphi_l \in X_\Delta \mid \varphi_l(P_j) = \delta_{lj},\ l = 1, \ldots, K;\ j = 1, \ldots, K\} \quad (5.1.7)\text{ (a)}$$

and a basis of $X_{\Delta,0}$ is:

$$B_{\Delta,0} = \{\varphi_l \in X_\Delta \mid \varphi_l(P_j) = \delta_{lj},\ l = 1, \ldots, M;\ j = 1, \ldots, K\} \quad (5.1.7)\text{ (b)}$$

(The Kronecker symbol δ_{lj} is defined by $\delta_{ll} = 1$ and $\delta_{lj} = 0$ for $l \neq j$). Every function $f_\Delta \in X_\Delta$ and $g_\Delta \in X_{\Delta,0}$ can be written as:

$$f_\Delta(x) = \sum_{l=1}^{K} f_l \varphi_l(x), \qquad x \in \bar{\Omega}, \qquad (5.1.8)\text{ (a)}$$

$$g_\Delta(x) = \sum_{l=1}^{M} g_l \varphi_l(x), \qquad x \in \bar{\Omega}, \qquad (5.1.8)\text{ (b)}$$

where $f_l = f_\Delta(P_l)$, $g_l = g_\Delta(P_l)$ are the values of f_Δ and g_Δ resp. at the vertex P_l. The vertices are the nodes of the finite element method.
We approximate the Dirichlet datum ψ_D, which is assumed to be continuous in $\partial\Omega_D$, by its piecewise linear interpolate $(\psi_D)_\Delta$ defined by:

$$(\psi_D)_\Delta := \sum_{l=1}^{K} \psi_D(P_l)\varphi_l. \qquad (5.1.9)$$

Clearly, the electron and hole concentrations n and p are normally not available exactly. We denote by n_Δ and p_Δ piecewise linear approximations of n and p obtained by discretising the continuity equations (as discussed in the next subsection):

$$n_\Delta = \sum_{l=1}^{K} n_l \varphi_l, \qquad p_\Delta = \sum_{l=1}^{K} p_l \varphi_l, \qquad (5.1.10)\text{ (a)}$$

$$n_l = n_\Delta(P_l), \qquad p_l = p_\Delta(P_l). \qquad (5.1.10)\text{ (b)}$$

Then the finite element approximation ψ_a of the potential ψ, which corresponds to the partition Δ, is the solution of the Ritz–Galerkin approximation (5.1.4) on the subspace $X_{\Delta,0}$ with piecewise linear Dirichlet datum $(\psi_D)_\Delta$ given by (5.1.9) and approximate space charge density $\varrho_a := -(n_\Delta - p_\Delta - C)$:

$$-\lambda^2 a(\psi_a, \varphi_\Delta) = (n_\Delta - p_\Delta - C, \varphi_\Delta)_{2,\Omega}$$

$$\text{for all } \varphi_\Delta \in X_{\Delta,0}, \qquad (5.1.11)\text{ (a)}$$

$$\psi_a - (\psi_D)_\Delta \in X_{\Delta,0}. \qquad (5.1.11)\text{ (b)}$$

The condition (5.1.11) (b) implies that ψ_a is piecewise linear on the mesh Δ, i.e. $\psi_a = \psi_\Delta \in X_\Delta$. Therefore it can be written in the form:

$$\psi_\Delta(x) = \sum_{l=1}^{K} \psi_l \varphi_l(x), \qquad \psi_l = \psi_\Delta(P_l). \tag{5.1.12}$$

By inserting (5.1.12) into (5.1.11), setting $\varphi_\Delta = \varphi_i$, $i = 1, \ldots, M$ and by using (5.1.9) we obtain the following system of equations for the nodal values ψ_j, $j = 1, \ldots, K$:

$$-\lambda^2 \sum_{l=1}^{K} \psi_l \int_\Omega \operatorname{grad} \varphi_l \cdot \operatorname{grad} \varphi_i \, dx = (n_\Delta - p_\Delta - C, \varphi_i)_{2,\Omega}$$

$$\text{for } i = 1, \ldots, M, \tag{5.1.13 (a)}$$

$$\psi_i = \psi_D(P_i), \qquad i = M+1, \ldots, K. \tag{5.1.13 (b)}$$

The right hand side of (5.1.13) (a) is usually evaluated by numerical integration. Let T denote an element of the partition Δ, $\mu_k(T)$ its k-dimensional Lebesguemeasure and let the function f be defined on T. A commonly used numerical integration formula is:

$$\int_T f(x) \, dx \approx \frac{\mu_k(T)}{k+1} \sum_{j=1}^{k+1} f_j =: M_T(f), \tag{5.1.14 (a)}$$

$$(f, g)_{2,\Omega,\Delta} := \sum_{l=1}^{N} M_{T_l}(fg), \tag{5.1.14 (b)}$$

where f_j denotes the value of f at the j-th vertex of T. In one space dimension (5.1.14) reduces to the trapezoidal rule.
The corresponding approximation of the right hand side of (5.1.13) (a) is given by:

$$(n_\Delta - p_\Delta - C, \varphi_i)_{2,\Omega,\Delta} = \frac{n_i - p_i - C_i}{k+1} \sum_{T_j \ni P_i} \mu_k(T_j). \tag{5.1.15}$$

The summation is performed over all elements, which have the vertex P_i. We denoted $C_i := C(P_i)$.
By inserting (5.1.15) into (5.1.13) (a) we obtain the discretisation:

$$-\lambda^2 \sum_{l=1}^{K} \psi_l \int_\Omega \operatorname{grad} \varphi_l \cdot \operatorname{grad} \varphi_i \, dx = \frac{n_i - p_i - C_i}{k+1} \sum_{T_j \ni P_i} \mu_k(T_j),$$

$$\text{for } \quad i = 1, \ldots, M, \tag{5.1.16 (a)}$$

$$\psi_i = \psi_D(P_i), \qquad i = M+1, \ldots, K. \tag{5.1.16 (b)}$$

Other finite element discretisations are obtained by appropriately defining higher order piecewise polynomials on simplicial elements (see, e.g., [5.30]) or by covering $\bar{\Omega}$ with different types of partitions consisting, for example, of quadrilateral (for $\Omega \subseteq \mathbb{R}^2$) or, resp., hexahedral (for $\Omega \subseteq \mathbb{R}^3$) elements.

C. Finite Difference Schemes

In the one-dimensional case the basis functions φ_i can be easily determined and a simple calculation reduces (5.1.16) to the three-point-discretisation:

$$\frac{2\lambda^2}{h_i+h_{i+1}}\left(\frac{\psi_{i+1}-\psi_i}{h_{i+1}}-\frac{\psi_i-\psi_{i-1}}{h_i}\right)=n_i-p_i-C_i, \tag{5.1.17 (a)}$$

for $\quad i=1,\ldots,N-1,$

$$\psi_0=\psi_D(-1), \qquad \psi_N=\psi_D(1), \tag{5.1.17 (b)}$$

where we denoted

$$h_i=x_i-x_{i-1}. \tag{5.1.18}$$

In multi-dimensional cases the finite element method (5.1.16) (approximately) coincides with commonly used finite difference methods for specially chosen partitions.

In the two-dimensional case we assume that the domain Ω is given by the interior of the rectangle with the corner points $(-a, 0)$, $(-a, b)$, (a, b), $(a, 0)$. We cover $\bar{\Omega}$ by a rectangular grid with the gridlines:

$$\begin{aligned} x&=x_i, \quad i=0,\ldots,N_x, \quad x_0=-a, \quad x_{N_x}=a;\\ y&=y_j, \quad j=0,\ldots,N_y, \quad y_0=0, \quad y_{N_y}=b \end{aligned} \tag{5.1.19 (a)}$$

and denote the grid spacings by:

$$h_i=x_i-x_{i-1}, \qquad k_j=y_j-y_{j-1}. \tag{5.1.19 (b)}$$

A finite element triangulation of $\bar{\Omega}$ is obtained by drawing the southwest-northeast diagonals of the grid-rectangles (see Fig. 5.1.2), i.e. by connecting the gridpoints (x_i, y_j) and (x_{i+1}, y_{j+1}), $i=0,\ldots,N_x-1$; $j=0,\ldots,N_y-1$. Then the left-hand side of (5.1.16) (a) is equal to $\lambda^2\frac{h_{i+1}+h_i}{2}\frac{k_{j+1}+k_j}{2}(\Delta_2\psi)_{ij}$, where $(\Delta_2\psi)_{ij}$ denotes the five-point finite difference approximation to the two-dimensional Laplacian of ψ evaluated at (x_i, y_j):

$$\begin{aligned}\Delta\psi(x_i,y_j)\approx(\Delta_2\psi)_{ij}:=&\frac{2}{h_{i+1}+h_i}\left(\frac{\psi_{i+1,j}-\psi_{ij}}{h_{i+1}}-\frac{\psi_{ij}-\psi_{i-1,j}}{h_i}\right)\\ &+\frac{2}{k_{j+1}+k_j}\left(\frac{\psi_{i,j+1}-\psi_{ij}}{k_{j+1}}-\frac{\psi_{i,j}-\psi_{i,j-1}}{k_j}\right).\end{aligned} \tag{5.1.20}$$

A typical five-point star is depicted in Fig. 5.1.3.

The standard five point discretisation of Poisson's equation reads:

$$(\Delta_2\psi)_{ij}=n_{ij}-p_{ij}-C_{ij}. \tag{5.1.21 (a)}$$

The difference formula (5.1.21) (a) only agrees with (5.1.16) (a), if the horizontal and the vertical gridlines are equidistant, i.e. $h_i\equiv h$, $k_j\equiv k$. Other-

wise terms involving mesh size differences in the right hand side of (5.1.16) (a) have to be dropped in order to obtain (5.1.21) (a).
The discrete Dirichlet data are given by:

$$\psi_{ij} = \psi_D(x_i, y_j), \qquad (x_i, y_j) \in \partial\Omega_D. \tag{5.1.21 (b)}$$

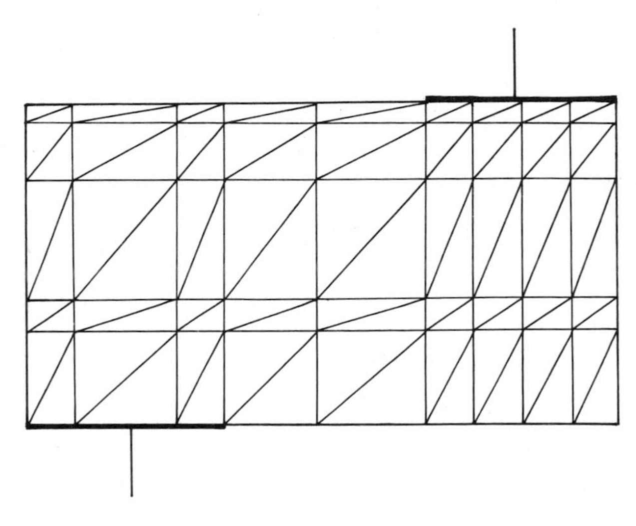

Fig. 5.1.2 Rectangular grid and associated finite element mesh

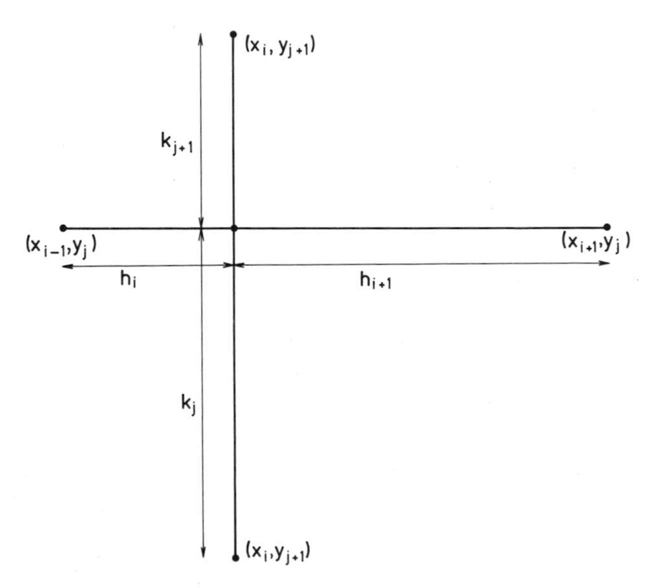

Fig. 5.1.3 A five point star

Usually, 'mirror imaging' is employed to discretise the Neumann conditions, i.e. (5.1.21) (a) is assumed to hold for $(x_i, y_j) \in \Omega \cup \partial\Omega_N$ and exterior grid-points (x_{-1}, y_j), (x_{N_x+1}, y_j) (x_i, y_{-1}), (x_i, y_{N_y+1}) with $h_0 = h_1$, $k_0 = k_1$, $h_{N_x+1} = h_{N_x}$, $k_{N_y+1} = k_{N_y}$ are introduced. Then the discrete Neumann conditions read:

$$\begin{pmatrix} \dfrac{\psi_{i+1,j} - \psi_{i-1,j}}{2h_i} \\ \dfrac{\psi_{i,j+1} - \psi_{i,j-1}}{2k_j} \end{pmatrix} \cdot \nu = \begin{pmatrix} 0 \\ 0 \end{pmatrix}, \qquad (x_i, y_j) \in \partial\Omega_N. \tag{5.1.21 (c)}$$

The advantage of 'mirror imaging' is that the local discretisation error on $\partial\Omega_N$ is of order two.

The difference equation (5.1.21) (a) can also be derived by the so called box integration method (see [5.32]).

If the device domain is not a rectangle, then the difference formula (5.1.21) (a) can be used at all interior grid points, only the discrete Neumann conditions have to be adjusted according to the geometry.

The derivation of three-dimensional finite difference schemes is straightforward.

Discretisation of the Current Relations and Continuity Equations

The steady state continuity equations for the current densities J_n and J_p read:

$$\left.\begin{aligned} \operatorname{div} J_n &= R \\ \operatorname{div} J_p &= -R \end{aligned}\right\} \quad x \in \Omega, \qquad \begin{aligned} &(5.1.22) \\ &(5.1.23) \end{aligned}$$

where $R = R(x, n, p)$ is the recombination-generation rate. We exclude avalanche generation for the following.

The current densities are given by the (scaled) current relations:

$$\left.\begin{aligned} J_n &= \mu_n(\operatorname{grad} n - n \operatorname{grad} \psi) \\ J_p &= -\mu_p(\operatorname{grad} p + p \operatorname{grad} \psi) \end{aligned}\right\} \quad x \in \Omega. \qquad \begin{aligned} &(5.1.24) \\ &(5.1.25) \end{aligned}$$

We insert (5.1.24), (5.1.25) into (5.1.22) and (5.1.23) resp. and obtain elliptic boundary value problems for n and p by taking into account the usual boundary conditions:

$$\operatorname{div}(\mu_n (\operatorname{grad} n - n \operatorname{grad} \psi)) = R, \qquad x \in \Omega \tag{5.1.26 (a)}$$

$$n|_{\partial\Omega_D} = n_D|_{\partial\Omega_D}, \qquad \left.\frac{\partial n}{\partial \nu}\right|_{\partial\Omega_N} = 0 \tag{5.1.26 (b)}$$

and

$$\operatorname{div}(\mu_p (\operatorname{grad} p + p \operatorname{grad} \psi)) = R, \qquad x \in \Omega \tag{5.1.27 (a)}$$

$$p|_{\partial\Omega_D} = p_D|_{\partial\Omega_D}, \qquad \left.\frac{\partial p}{\partial \nu}\right|_{\partial\Omega_N} = 0. \tag{5.1.27 (b)}$$

The Dirichlet data n_D, p_D are assumed to be in $H^1(\Omega)$, their restrictions to $\partial\Omega_D$ in $C(\partial\Omega_D)$ and $n_D|_{\partial\Omega_D}, p_D|_{\partial\Omega_D} > 0$.

A. Weak Formulations

The weak formulation of (5.1.26) reads

$$-b_n(\psi; n, \varphi) = (R, \varphi)_{2,\Omega} \quad \text{for all } \varphi \in H_0^1(\Omega \cup \partial\Omega_N), \tag{5.1.28 (a)}$$

$$n - n_D \in H_0^1(\Omega \cup \partial\Omega_N) \tag{5.1.28 (b)}$$

and the weak formulation of (5.1.27):

$$-b_p(\psi; p, \varphi) = (R, \varphi)_{2,\Omega} \quad \text{for all } \varphi \in H_0^1(\Omega \cup \partial\Omega_N), \tag{5.1.29 (a)}$$

$$p - p_D \in H_0^1(\Omega \cup \partial\Omega_N). \tag{5.1.29 (b)}$$

Here $b_n(\psi; .,.)$ and $b_p(\psi; .,.)$ denote the bilinear forms:

$$\begin{aligned} b_n(\psi; n, \varphi) &:= \int_\Omega J_n \cdot \operatorname{grad} \varphi \, dx \\ &= \int_\Omega \mu_n(\operatorname{grad} n \cdot \operatorname{grad} \varphi - n \operatorname{grad} \psi \cdot \operatorname{grad} \varphi) \, dx, \end{aligned} \tag{5.1.30}$$

$$\begin{aligned} b_p(\psi; p, \varphi) &:= -\int_\Omega J_p \cdot \operatorname{grad} \varphi \, dx \\ &= \int_\Omega \mu_p(\operatorname{grad} p \cdot \operatorname{grad} \varphi + p \operatorname{grad} \psi \cdot \operatorname{grad} \varphi) \, dx, \end{aligned} \tag{5.1.31}$$

which are defined for all $\varphi \in H^1(\Omega)$, $n \in H^1(\Omega) \cap L^\infty(\Omega)$ and $p \in H^1(\Omega) \cap L^\infty(\Omega)$ resp., if $\psi \in H^1(\Omega)$ and if μ_n, μ_p are bounded.
In the sequel we shall – as for the analysis of the static device problem presented in Chapter 3 – employ the set of dependent variables (ψ, u, v) defined by:

$$n = \delta^2 e^{\psi} u, \qquad p = \delta^2 e^{-\psi} v. \tag{5.1.32}$$

The current relations (5.1.24), (5.1.25) then read:

$$J_n = \delta^2 \mu_n e^{\psi} \operatorname{grad} u, \tag{5.1.33 (a)}$$

$$J_p = -\delta^2 \mu_p e^{-\psi} \operatorname{grad} v \tag{5.1.33 (b)}$$

and the strong formulations of the boundary value problems (5.1.26), (5.1.27) reduce to:

$$\operatorname{div}(\delta^2 \mu_n e^{\psi} \operatorname{grad} u) = R, \qquad x \in \Omega \tag{5.1.34 (a)}$$

$$u|_{\partial\Omega_D} = u_D|_{\partial\Omega_D}, \quad \left.\frac{\partial u}{\partial \nu}\right|_{\partial\Omega_N} = 0, \tag{5.1.34 (b)}$$

$$\operatorname{div}(\delta^2 \mu_p e^{-\psi} \operatorname{grad} v) = R, \qquad x \in \Omega \tag{5.1.35 (a)}$$

$$v|_{\partial\Omega_D} = v_D|_{\partial\Omega_D}, \quad \left.\frac{\partial v}{\partial \nu}\right|_{\partial\Omega_N} = 0 \tag{5.1.35 (b)}$$

with $u_D = \frac{1}{\delta^2} e^{-\psi_D} n_D$, $v_D = \frac{1}{\delta^2} e^{\psi_D} p_D$. The weak formulations are given by:

$$-b_u(\psi; u, \varphi) = (R, \varphi)_{2,\Omega} \quad \text{for all } \varphi \in H_0^1(\Omega \cup \partial\Omega_N), \tag{5.1.36 (a)}$$

$$u - u_D \in H_0^1(\Omega \cup \partial\Omega_N), \tag{5.1.36 (b)}$$

$$-b_v(\psi; v, \varphi) = (R, \varphi)_{2,\Omega} \quad \text{for all } \varphi \in H_0^1(\Omega \cup \partial\Omega_N), \tag{5.1.37 (a)}$$

$$v - v_D \in H_0^1(\Omega \cup \partial\Omega_N), \tag{5.1.37 (b)}$$

where the bilinear forms

$$b_u(\psi; u, \varphi) := \int_\Omega \delta^2 \mu_n e^{\psi} \operatorname{grad} u \cdot \operatorname{grad} \varphi \, dx \tag{5.1.38}$$

$$b_v(\psi; v, \varphi) := \int_\Omega \delta^2 \mu_p e^{-\psi} \operatorname{grad} v \cdot \operatorname{grad} \varphi \, dx \tag{5.1.39}$$

relate to b_n and b_p resp. in the following way:

$$b_u(\psi; u, \varphi) = b_n(\psi; \delta^2 e^{\psi} u, \varphi), \tag{5.1.40 (a)}$$

$$b_v(\psi; v, \varphi) = b_p(\psi; \delta^2 e^{-\psi} v, \varphi). \tag{5.1.40 (b)}$$

B. The Failure of the Standard Discretisation

It is well known to device modelers that the discretisation of the current relations requires particular care. To demonstrate the reason for this, we examine a one-dimensional steeply graded *pn*-junction diode in thermal equilibrium. We set $R \equiv 0$ and $\mu_n \equiv 1$. Then (5.1.26) reads:

$$(n' - n\psi_e')' = 0, \quad -1 \leq x \leq 1, \tag{5.1.41 (a)}$$

$$n(-1) = n_D(-1), \quad n(1) = n_D(1). \tag{5.1.41 (b)}$$

In Chapter 4 we demonstrated that a steeply graded junction at $x = X \in (-1, 1)$ produces a thin layer of fast variation in the potential and in the carrier concentration at $x = X$. We depict a typical equilibrium potential $\psi = \psi_e$ in Fig. 5.1.4. Assuming that the contacts at $x = \pm 1$ are Ohmic, the Dirichlet data satisfy $n_D(\pm 1) = \delta^2 e^{\psi_e(\pm 1)}$ and the solution of (5.1.41) is given by $n(x) = n_e(x) = \delta^2 e^{\psi_e(x)}$ (see Fig. 5.1.4).
We set up the following straightforward trapezoidal-type difference scheme:

$$J_{n,i+\frac{1}{2}} := \frac{n_{i+1} - n_i}{h_{i+1}} - \frac{n_{i+1} + n_i}{2} \frac{\psi_{i+1} - \psi_i}{h_{i+1}}, \tag{5.1.42 (a)}$$

$$\frac{2}{h_i + h_{i+1}} (J_{n,i+\frac{1}{2}} - J_{n,i-\frac{1}{2}}) = 0 \quad \text{for } i = 1, \ldots, N-1, \tag{5.1.42 (b)}$$

$$n_0 = n_D(-1), \quad n_N = n_D(1). \tag{5.1.42 (c)}$$

$J_{n,i+\frac{1}{2}}$ is regarded as approximation of $J_n(x_{i+\frac{1}{2}})$, where $x_{i+\frac{1}{2}} := x_i + \frac{h_{i+1}}{2}$.

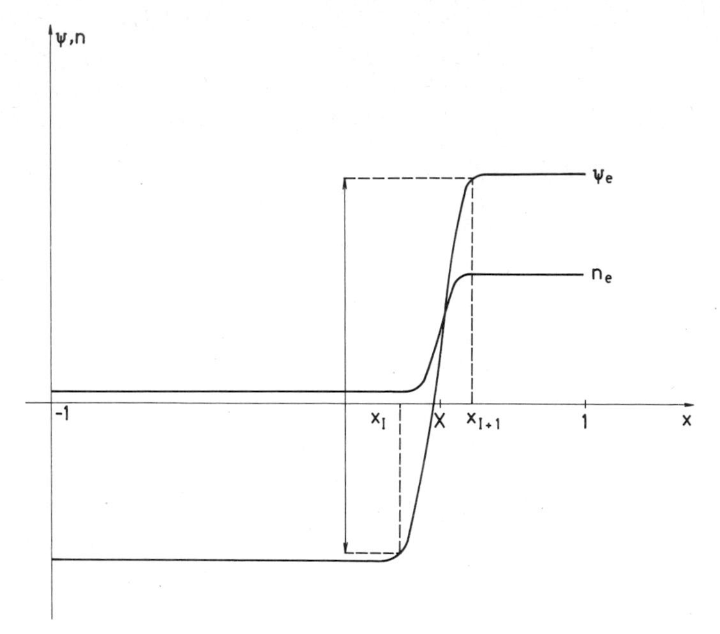

Fig. 5.1.4 Equilibrium potential and electron concentration of a *PN*-diode

(5.1.42) (b) implies that $J_{n,i+\frac{1}{2}}$ is constant, i.e. $J_{n,i+\frac{1}{2}} \equiv \tilde{J}_n$ and from (5.1.42) (a) we obtain:

$$n_{i+1} = \frac{1+\frac{z_i}{2}}{1-\frac{z_i}{2}} n_i + \frac{h_{i+1}}{1-\frac{z_i}{2}} \tilde{J}_n, \qquad z_i := \psi_{i+1} - \psi_i. \tag{5.1.43}$$

Assume now that the discrete potential coincides with the equilibrium potential, i.e. $\psi_i = \psi_e(x_i)$, that $\tilde{J}_n$ approximates the equilibrium electron current density $J_{n_e} \equiv 0$, i.e. $\tilde{J}_n \approx 0$, and that for some index $i = I$ the discrete solution n_I approximates the exact solution $n_e(x_I)$, i.e. $n_I \approx \delta^2 e^{\psi_e(x_I)}$. Then, if the grid is such that $\psi_e(x_{i+1}) - \psi_e(x_i)$ is bounded away from 2, we obtain from (5.1.43):

$$n_{I+1} \approx \frac{1+\frac{z_I}{2}}{1-\frac{z_I}{2}} \delta^2 e^{\psi_e(x_I)}$$

and n_{I+1} approximates $n_e(x_{I+1}) = \delta^2 e^{\psi_e(x_{I+1})}$ if and only if $(1+(z_I/2))/(1-(z_I/2))$ approximates $e^{\psi_e(x_{I+1})-\psi_e(x_I)}$, that is if and only if $(1+(z_I/2))/(1-(z_I/2)) \approx e^{z_I}$. Since $(1+(z/2))/(1-(z/2))$ only approximates e^z if z is small, we conclude that the solutions n_i, $J_{n,i+(1/2)}$ of the scheme (5.1.42) only approximate the exact solutions $n_e(x_i)$, J_{n_e} at every gridpoint, if the potential is resolved accurately by the mesh, i.e. if it varies only weakly on each subinterval:

$$\max_i |\psi_{i+1} - \psi_i| \ll 1. \qquad (5.1.44)$$

This is a very restrictive condition on the mesh, particularly since it requires that small meshsizes are used inside the junction layer. If, for example, no gridpoint is placed inside the layer (see Fig. 5.1.4), then $|\psi_{I+1} - \psi_I|$ is large, and the scheme (5.1.42) does not provide reasonable approximations. Particularly for multidimensional problems the restriction that the potential only varies weakly on each of the finite elements may require extremly many gridpoints and is therefore very costly and in some cases even impossible to guarantee. Therefore the simple scheme (5.1.42) and its multidimensional analogues are practically not feasable.

C. Finite Element Scharfetter-Gummel Discretisations

Appropriate discretisations of the one-dimensional current relations can be derived by using that – for $k = 1$ – the current densities J_n and J_p are moderately varying functions, i.e. they do not exhibit layer behaviour. Thus they are well approximated by a constant locally (see [5.22], [5.28] for a derivation of the schemes). In multidimensional cases the current densities may exhibit – strictly speaking – layer behaviour (see Chapter 4), however, they very often vary slower than ψ, n and p.
We shall now derive finite element discretisations of the current relations, which are based on this observation (see [5.4], [5.35]).
Since we shall have to perform various computations on each of the elements of the partition, it is advantageous to transform the elements into a master (reference) element T_m. Therefore, let $P_1, \ldots, P_{k+1}$ be the vertices of the k-dimensional simplicial element T. Then the coordinate transformation:

$$x = x(\xi) := \sum_{j=1}^{k+1} P_j M_j(\xi), \qquad x = (x_1, \ldots, x_k), \quad \xi = (\xi_1, \ldots, \xi_k) \qquad (5.1.45)\ (a)$$

with the shape-functions

$$M_1(\xi) = 1 - \xi_1 - \ldots - \xi_k, \qquad M_2(\xi) = \xi_1, \ldots, M_{k+1}(\xi) = \xi_k \qquad (5.1.45)\ (b)$$

transforms the master interval [0, 1] into T for $k = 1$, the master triangle with the vertices (0, 0), (1, 0), (0, 1) into the triangle T for $k = 2$ and, resp., the master tetrahedron with the vertices (0, 0, 0), (1, 0, 0), (0, 1, 0), (0, 0, 1) into the tetrahedron T for $k = 3$. For a function f defined on T we set:

$$f^T(\xi) := f(x(\xi)), \quad \xi \in T_m. \qquad (5.1.46)\ (a)$$

A piecewise linear function $f_\Delta \in X_\Delta$ satisfies:

$$f_\Delta^T(\xi) = f_\Delta(x)|_T = \sum_{j=1}^{k+1} f_j M_j(\xi), \quad \xi \in T_m, \qquad (5.1.46)\ (b)$$

where f_j is the value of f at the vertex P_j of T.

We denote by G_m the center of gravity of the reference element and by $x_{G_T} = x(G_m)$ the image of G_m in the element T.
We obtain an approximation of the current relation (5.1.33) (a) in the element T by approximating μ_n by its value μ_{n,G_T} at x_{G_T}, ψ by a piecewise linear function ψ_Δ and by approximating J_n by a constant vector J_{n,G_T}:

$$J_{n,G_T} = \delta^2 \mu_{n,G_T} e^{\psi_\Delta(x)} \operatorname{grad} u_a(x), \qquad x \in T. \tag{5.1.47) (a}$$

Applying the coordinate transformation (5.1.45) gives

$$J_{n,G_T} = \delta^2 \mu_{n,G_T} e^{\psi_\Delta^T(\xi)} (I'_T)^{-1} \operatorname{grad}_\xi u_a^T(\xi), \qquad \xi \in T_m, \tag{5.1.47) (b}$$

where I_T denotes the Jacobian of the map $x = x(\xi)$:

$$I_T = \frac{\partial x}{\partial \xi} = \left(\frac{\partial x_i}{\partial \xi_j}\right)_{\substack{i=1,\dots,k\\ j=1,\dots,k}} \tag{5.1.47) (c}$$

(and I'_T its transposed).
We multiply (5.1.47) (b) by $e^{-\psi_\Delta^T(\xi)} I'_T$, integrate the j-th component with respect to ξ_j in the interval [0, 1], observe that $\dfrac{\partial x}{\partial \xi}$ is constant in T_m and obtain after transformation to the x-coordinates:

$$J_{n,G_T} = \delta^2 \mu_{n,G_T} e^{\psi_\Delta(P_1)} (I'_T)^{-1} B_T(\psi_\Delta) I'_T \operatorname{grad} u_\Delta|_T, \tag{5.1.48) (a}$$

where u_Δ is the piecewise linear interpolate of u_a. $B_T(\psi_\Delta)$ denotes the diagonal matrix:

$$B_T(\psi_\Delta) = \operatorname{diag}\left(B(\psi_\Delta(P_1) - \psi_\Delta(P_2)), \dots, B(\psi_\Delta(P_1) - \psi_\Delta(P_{k+1}))\right), \tag{5.1.48) (b}$$

and B the Bernoulli function

$$B(z) := \frac{z}{e^z - 1}. \tag{5.1.48) (c}$$

By proceeding analogously with the hole current relation (5.1.33) (b) we derive:

$$J_{p,G_T} = -\delta^2 \mu_{p,G_T} e^{-\psi_\Delta(P_1)} (I'_T)^{-1} D_T(\psi_\Delta) I'_T \operatorname{grad} v_\Delta|_T \tag{5.1.49) (a}$$

with $D_T(\psi_\Delta)$ given by:

$$D_T(\psi_\Delta) = B_T(-\psi_\Delta). \tag{5.1.49) (b}$$

We define

$$F_T(\psi_\Delta) := e^{\psi_\Delta(P_1)} (I'_T)^{-1} B_T(\psi_\Delta) I'_T, \tag{5.1.50) (a}$$

$$H_T(\psi_\Delta) := e^{-\psi_\Delta(P_1)} (I'_T)^{-1} D_T(\psi_\Delta) I'_T \tag{5.1.50) (b}$$

and rewrite (5.1.48) (a), (5.1.49) (a) as

$$J_{n,G_T} = \delta^2 \mu_{n,G_T} F_T(\psi_\Delta) \operatorname{grad} u_\Delta|_T \tag{5.1.50) (c}$$

$$J_{p,G_T} = -\delta^2\mu_{p,G_T} H_T(\psi_\Delta) \operatorname{grad} v_\Delta|_T. \tag{5.1.50 (d)}$$

The matrices $F_T(\psi_\Delta)$, $H_T(\psi_\Delta)$ may depend on the choice of the node P_1 of the element T. In order to remove this arbitrariness from the definition of the discrete combined current relations-continuity equations we proceed as follows. Let P_i be an interior or Neumann-boundary node. Then, for all elements, which have P_i as vertex, we choose the transformations (5.1.45) such that P_i is mapped into 0, (0, 0) or (0, 0, 0), resp. Consequently, for all those elements T, the matrices F_T, H_T depend on P_i, i.e. $F_T(\psi_\Delta) = F_{T,i}(\psi_\Delta)$, $H_T(\psi_\Delta) = H_{T,i}(\psi_\Delta)$. Since the basis function φ_i, which 'corresponds' to P_i, vanishes in all elements, which do not have P_i as vertex, we may define discrete approximations of $b_u(\psi, u, \varphi_i)$ and $b_v(\psi, v, \varphi_i)$ (the quadratic forms b_u, b_v are defined in (5.1.38), (5.1.39)) as:

$$b_{u,i}(\psi_\Delta; u_\Delta, \varphi_i) := \sum_{T_j \ni P_i} \delta^2\mu_{n,G_{T_j}} \int_{T_j} F_{T_{j,i}}(\psi_\Delta) \operatorname{grad} u_\Delta \cdot \operatorname{grad} \varphi_i \, dx \tag{5.1.51 (a)}$$

$$b_{v,i}(\psi_\Delta; v_\Delta, \varphi_i) := \sum_{T_j \ni P_i} \delta^2\mu_{p,G_{T_j}} \int_{T_j} H_{T_{j,i}}(\psi_\Delta) \operatorname{grad} v_\Delta \cdot \operatorname{grad} \varphi_i \, dx \tag{5.1.51 (b)}$$

For a piecewise linear function $\varphi_\Delta \in X_\Delta$ of the form

$$\varphi_\Delta(x) = \sum_{l=1}^{K} \varphi_\Delta(P_l)\varphi_l$$

we define the discrete analogues of b_u, b_v:

$$b_{u,\Delta}(\psi_\Delta, u_\Delta, \varphi_\Delta) := \sum_{l=1}^{K} \varphi_\Delta(P_l) b_{u,l}(\psi_\Delta; u_\Delta, \varphi_l) \tag{5.1.52 (a)}$$

$$b_{v,\Delta}(\psi_\Delta, v_\Delta, \varphi_\Delta) := \sum_{l=1}^{K} \varphi_\Delta(P_l) b_{v,l}(\psi_\Delta; v_\Delta, \varphi_l). \tag{5.1.52 (b)}$$

The quadratic forms $b_{u,\Delta}(\psi_\Delta; .,.)$ and $b_{v,\Delta}(\psi_\Delta; .,.)$ are independent of the element transformations and symmetric on X_Δ (see [5.35]).

In the one-dimensional case it is easy to show that $F_T(\psi_\Delta)$, $H_T(\psi_\Delta)$ are independent of the numbering of the end-points of the interval T. Thus, for $k=1$, the discrete current densities J_{n,G_T}, J_{p,G_T} are uniquely defined. In two or three space dimensions J_{n,G_T} and J_{p,G_T} depend on the numbering of the vertrices of the element T, i.e. the discrete current densities in T are not uniquely defined for $k = 2, 3$.

The finite element discretisations of the combined current relations-continuity equations read:

$$-b_{u,\Delta}(\psi_\Delta; u_\Delta, \varphi_\Delta) = (R, \varphi_\Delta)_{2,\Omega,\Delta}$$

$$\text{for all } \varphi_\Delta \in X_{\Delta,0}, \tag{5.1.53 (a)}$$

$$u_\Delta|_{\partial\Omega_D} = (u_D)_\Delta|_{\partial\Omega_D} \tag{5.1.53 (b)}$$

and, resp.:

$$-b_{v,\Delta}(\psi_\Delta; v_\Delta, \varphi_\Delta) = (R, \varphi_\Delta)_{2,\Omega,\Delta}$$

$$\text{for all } \varphi_\Delta \in X_{\Delta,0}, \tag{5.1.54) (a}$$

$$v_\Delta|_{\partial\Omega_D} = (v_D)_\Delta|_{\partial\Omega_D}. \tag{5.1.54) (b}$$

The discrete current relations can be expressed in terms of the n, p-variables. We define the piecewise linear functions $n_\Delta, p_\Delta \in X_\Delta$ by their nodal values:

$$n_\Delta(P_l) := \delta^2 e^{\psi_\Delta(P_l)} u_\Delta(P_l), \qquad l = 1, \ldots, K \tag{5.1.55) (a}$$

$$p_\Delta(P_l) := \delta^2 e^{-\psi_\Delta(P_l)} v_\Delta(P_l), \qquad l = 1, \ldots, K, \tag{5.1.55) (b}$$

and obtain for the element T with the vertices $P_1, \ldots, P_{k+1}$:

$$J_{n,G_T} = \mu_{n,G_T}\left(\Gamma_T^n(\psi_\Delta)\,\mathrm{grad}\, n_\Delta - \frac{n_\Delta(P_1)+\ldots+n_\Delta(P_{k+1})}{k+1}\,\mathrm{grad}\,\psi_\Delta\right), \qquad x \in T \tag{5.1.56}$$

$$J_{p,G_T} = -\mu_{p,G_T}\left(\Gamma_T^p(\psi_\Delta)\,\mathrm{grad}\, p_\Delta + \frac{p_\Delta(P_1)+\ldots+p_\Delta(P_{k+1})}{k+1}\,\mathrm{grad}\,\psi_\Delta\right), \qquad x \in T, \tag{5.1.57}$$

where the $k \times k$-matrices $\Gamma_T^n(\psi_\Delta)$, $\Gamma_T^p(\psi_\Delta)$ are defined by:

$$\Gamma_T^n(\psi_\Delta) := (I_T')^{-1}\left[\frac{1}{k+1}\begin{pmatrix}\psi_\Delta(P_2)-\psi_\Delta(P_1)\\ \vdots \\ \psi_\Delta(P_{k+1})-\psi_\Delta(P_1)\end{pmatrix}(1, \ldots, 1) + e^{\psi_\Delta(P_1)} B_T(\psi_\Delta)\,\mathrm{diag}\,(e^{-\psi_\Delta(P_2)}, \ldots, e^{-\psi_\Delta(P_{k+1})})\right] I_T' \tag{5.1.58) (a}$$

and

$$\Gamma_T^p(\psi_\Delta) := \Gamma_T^n(-\psi_\Delta). \tag{5.1.58) (b}$$

Because of the introduced numbering of the vertices the matrices Γ_T^n and Γ_T^p also depend on the node P_i for $k>1$, i.e. $\Gamma_T^n(\psi_\Delta) = \Gamma_{T,i}^p(\psi_\Delta)$, $\Gamma_T^p(\psi_\Delta) = \Gamma_{T,i}^p(-\psi_\Delta)$ (for the elements, which have the vertex P_i).
Since $\int_T f_\Delta\,dx = (\mu_k(T)/(k+1))\,(f_\Delta(P_1)+\ldots+f_\Delta(P_{k+1}))$ for every $f_\Delta \in X_\Delta$ holds, we set

$$b_{n,i}(\psi_\Delta; n_\Delta, \varphi_i) := \sum_{T_j \ni P_i} \mu_{n,G_{T_j}} \int_{T_j} (\Gamma_{T_j,i}^n(\psi_\Delta)\,\mathrm{grad}\, n_\Delta \cdot \mathrm{grad}\,\varphi_i - n_\Delta\,\mathrm{grad}\,\psi_\Delta \cdot \mathrm{grad}\,\varphi_i)\,dx \tag{5.1.59) (a}$$

$$b_{p,i}(\psi_\Delta; p_\Delta, \varphi_i) := \sum_{T_j \ni P_i} \mu_{p,G_{T_j}} \int_{T_j} (\Gamma^p_{T_j,i}(\psi_\Delta) \operatorname{grad} p_\Delta \cdot \operatorname{grad} \varphi_i + p_\Delta \operatorname{grad} \psi_\Delta \cdot \operatorname{grad} \varphi_i)\, dx. \qquad (5.1.59)\ (b)$$

Then, for $\varphi_\Delta \in X_\Delta$, the discrete analogues of the bilinear forms b_n and b_p of (5.1.30) and (5.1.31) resp. read:

$$b_{n,\Delta}(\psi_\Delta; n_\Delta, \varphi_\Delta) := \sum_{l=1}^{K} \varphi_\Delta(P_l) b_{n,l}(\psi_\Delta; n_\Delta, \varphi_l) \qquad (5.1.60)\ (a)$$

$$b_{p,\Delta}(\psi_\Delta; p_\Delta, \varphi_\Delta) := \sum_{l=1}^{K} \varphi_\Delta(P_l) b_{p,l}(\psi_\Delta; p_\Delta, \varphi_l). \qquad (5.1.60)\ (b)$$

The finite element discretisation of (5.1.28) is given by:

$$-b_{n,\Delta}(\psi_\Delta; n_\Delta, \varphi_\Delta) = (R, \varphi_\Delta)_{2,\Omega,\Delta}$$

$$\text{for all } \varphi_\Delta \in X_{\Delta,0}, \qquad (5.1.61)\ (a)$$

$$n_\Delta|_{\partial\Omega_D} = (n_D)_\Delta|_{\partial\Omega_D} \qquad (5.1.61)\ (b)$$

and the discretisation of (5.1.29):

$$-b_{p,\Delta}(\psi_\Delta; p_\Delta, \varphi_\Delta) = (R, \varphi_\Delta)_{2,\Omega,\Delta}$$

$$\text{for all } \varphi_\Delta \in X_{\Delta,0}, \qquad (5.1.62)\ (a)$$

$$p_\Delta|_{\partial\Omega_D} = (p_D)_\Delta|_{\partial\Omega_D}. \qquad (5.1.62)\ (b)$$

Systems of equations for the nodal values $n_l := n_\Delta(P_l)$, $p_l := p_\Delta(P_l)$ are obtained by inserting the expressions:

$$n_\Delta = \sum_{l=1}^{K} n_l \varphi_l, \qquad p_\Delta = \sum_{l=1}^{K} p_l \varphi_l, \qquad \psi_\Delta = \sum_{l=1}^{K} \psi_l \varphi_l \qquad (5.1.63)$$

into (5.1.61), (5.1.62) and setting $\varphi_\Delta = \varphi_i$, $i = 1, \ldots, M$. Then (5.1.61) reads:

$$-\sum_{T_j \ni P_i} \mu_{n,G_{T_j}} \left[\sum_{l=1}^{K} n_l \int_{T_j} \Gamma^n_{T_j,l}(\psi_\Delta) \operatorname{grad} \varphi_l \cdot \operatorname{grad} \varphi_i \, dx - \frac{\sum_{P_m \in T_j} n_m}{k+1} \sum_{l=1}^{K} \psi_l \int_{T_j} \operatorname{grad} \varphi_l \cdot \operatorname{grad} \varphi_i \, dx \right] = \frac{R_i}{k+1} \sum_{T_j \ni P_i} \mu_k(T_j)$$

$$\text{for } i = 1, \ldots, M, \qquad (5.1.64)\ (a)$$

$$n_i = n_D(P_i), \qquad i = M+1, \ldots, K. \qquad (5.1.64)\ (b)$$

If $R = R(x, n, p)$ and $\mu_{n,p} = \mu_{n,p}(x, \operatorname{grad} \psi)$, then we set:

$$R_i := R(P_i, n_i, p_i), \qquad (5.1.65)\ (a)$$

$$\mu_{n,G_{T_i}} := \mu_n(x_{G_{T_i}}, \operatorname{grad} \psi_\Delta|_{T_i}), \qquad (5.1.65)\ (b)$$

$$\mu_{p,G_{T_i}} := \mu_p(x_{G_{T_i}}, \operatorname{grad} \psi_\Delta|_{T_i}), \qquad (5.1.65)\ (c)$$

(grad ψ_Δ is constant on each element!).

In the one-dimensional case $\Gamma^n_{T_i}$ is given by

$$\Gamma^n_{T_i}(\psi_\Delta) = \gamma(\psi_{i+1} - \psi_i), \qquad \gamma(z) := \frac{z}{2}\,\frac{e^z+1}{e^z-1}. \tag{5.1.66}$$

The one-dimensional versions of (5.1.56), (5.1.57) were suggested first by Scharfetter and Gummel [5.28]. We shall therefore also refer to the finite element discretisations as Scharfetter–Gummel schemes.
The multipler $\gamma(\psi_{i+1} - \psi_i)$ distinguishes the Scharfetter–Gummel schemes (called SG-schemes in the sequel) from the 'straightforward' discretisation (5.1.42). Since $\gamma(z) = 1 + O(z^2)$ as $z \to 0$, the SG-scheme is approximately equivalent to (5.1.42) for those meshes, which allow an accurate resolution of the discrete potential. The factor $\gamma(\psi_{i+1} - \psi_i)$ is decisively important, if the mesh does not resolve the potential well. The underlying reason is that the difference scheme

$$\gamma(\psi(x_{i+1}) - \psi(x_i))\,\frac{n_{i+1} - n_i}{h_{i+1}} = \frac{\psi(x_{i+1}) - \psi(x_i)}{h_{i+1}}\,\frac{n_{i+1} + n_i}{2},$$

$$n_0 = n_D(-1)$$

integrates the initial value problem

$$n' = \psi' n, \qquad n(-1) = n_D(-1)$$

exactly for all functions ψ'. In the terminology of [5.8] γ is called an exponential fitting function. The SG-discretisations are exponentially fitted schemes for the current relations.

D. The Finite Difference Scharfetter–Gummel Schemes

Let us now assume that – for a rectangular two-dimensional domain Ω – the triangulation Δ originates from the rectangular grid given by (5.1.19). Then the discretisations (5.1.53), (5.1.54) (in the u, v-variables) and (5.1.61), (5.1.62) (in the n, p-variables) reduce to five-point finite difference schemes. By appropriately modifying the right hand sides and the equations corresponding to Neumann-boundary points and by evaluating the mobilities at the interval midpoints instead of at the centers of gravity of the finite elements we obtain the commonly used two-dimensional SG-finite difference discretisations of (5.1.34):

$$J^x_{n,i+\frac{1}{2},j} := \delta^2 \mu_{n,i+\frac{1}{2},j}\, e^{\psi_{i+1,j}} B(\psi_{i+1,j} - \psi_{ij})\,\frac{u_{i+1,j} - u_{ij}}{h_{i+1}}$$

$$J^y_{n,i,j+\frac{1}{2}} := \delta^2 \mu_{n,i,j+\frac{1}{2}}\, e^{\psi_{i,j+1}} B(\psi_{i,j+1} - \psi_{ij})\,\frac{u_{i,j+1} - u_{ij}}{k_{j+1}} \tag{5.1.67 (a)}$$

$$\frac{2}{h_{i+1}+h_i}\left(J^x_{n,i+\frac{1}{2},j} - J^x_{n,i-\frac{1}{2},j}\right) + \frac{2}{k_{j+1}+k_j}\left(J^y_{n,i,j+\frac{1}{2}} - J^y_{n,i,j-\frac{1}{2}}\right) = R_{ij}$$

for all $(x_i, y_j) \in \Omega \cup \partial\Omega_N$, (5.1.67) (b)

$$h_{N_x+1} := h_{N_x}, \quad h_0 := h_1, \quad k_{N_y+1} := k_{N_y}, \quad k_0 := k_1 \qquad (5.1.67)\ (c)$$

$$u_{ij} = u_D(x_i, y_j), \quad (x_i, y_j) \in \partial\Omega_D \qquad (5.1.67)\ (d)$$

$$\begin{pmatrix} \dfrac{u_{i+1,j} - u_{i-1,j}}{h_i} \\ \dfrac{u_{i,j+1} - u_{i,j-1}}{k_j} \end{pmatrix} \cdot v = 0, \quad (x_i, y_j) \in \partial\Omega_N. \qquad (5.1.67)\ (e)$$

The finite difference equations can easily be rewritten in terms of

$$n_{ij} = \delta^2 e^{\psi_{ij}} u_{ij}, \quad p_{ij} = \delta^2 e^{-\psi_{ij}} v_{ij}, \qquad (5.1.68)$$

e.g. the discrete y-component of the electron current density reads:

$$J^y_{n,i,j+\frac{1}{2}} = \mu_{n,i,j+\frac{1}{2}} \left(\gamma(\psi_{i,j+1} - \psi_{ij}) \frac{n_{i,j+1} - n_{ij}}{k_{j+1}} - \frac{n_{i,j+1} + n_{ij}}{2} \frac{\psi_{i,j+1} - \psi_{ij}}{k_{j+1}} \right). \qquad (5.1.69)$$

The SG-finite difference schemes can be extended to three dimensions in a straightforward way.
Mock [5.23], [5.24] analysed similar SG-type finite difference methods on rectangular and triangular grids. Zlamal [5.35] derived SG-type discretisations also on quadrilateral and hexahedral elements using piecewise bilinear and, resp., piecewise trilinear test-functions.

5.2 Solutions of Static Discretisations

The piecewise linear finite element discretisation of the static semiconductor device problem consists of the discrete Poisson's equation (5.1.16) and the discrete current relation-continuity equations (5.1.61), (5.1.62). We assume an appropriate ordering of mesh-points (see [5.29] for various possibilities) and denote the vectors of node-values by ψ^Δ, n^Δ, p^Δ and C^Δ resp. Then (5.1.16) can be written as:

$$(SDFE1) \quad -\lambda^2 A(\Delta)\psi^\Delta = D(\Delta)\,(n^\Delta - p^\Delta - C^\Delta) + \psi^\Delta_D$$

and (5.1.61), (5.1.62) read:

$$(SDFE2) \quad -B_n(\Delta; \psi^\Delta) n^\Delta = D(\Delta) R(\Delta; n^\Delta, p^\Delta) + n^\Delta_D(\psi^\Delta)$$

$$(SDFE3) \quad -B_p(\Delta; \psi^\Delta) p^\Delta = D(\Delta) R(\Delta; n^\Delta, p^\Delta) + p^\Delta_D(\psi^\Delta).$$

A, D, B_n and B_p are $M \times M$-matrices; R and ψ^Δ_D, n^Δ_D, p^Δ_D are M-vectors representing the discretised recombination-generation rate $R(x, n, p)$ and the contributions steming from the Dirichlet boundary conditions resp. The i-th entry of the diagonal matrix D is given by $(1/(k+1)) \sum_{T_l \ni P_i} \mu_k(T_l)$.

The problem *(SDFE)* constitutes a nonlinearly coupled system of algebraic equations for the nodal values ψ_i, n_i and p_i.
We shall also use the formulation of the discrete system in terms of the variables (ψ_i, u_i, v_i) $(n_i = \delta^2 e^{\psi_i} u_i,\ p_i = \delta^2 e^{-\psi_i} v_i)$ given by (5.1.16), (5.1.53), (5.1.54). By assembling the equations we obtain a nonlinear system of the following form:

$$(SDFEM1) \quad -\lambda^2 A(\Delta)\psi^\Delta = D(\Delta)\,(\delta^2 E(\psi^\Delta)u^\Delta - \delta^2 E^{-1}(\psi^\Delta)v^\Delta - C^\Delta) + \psi_D^\Delta$$

$$(SDFEM2) \quad -B_u(\Delta;\psi^\Delta)u^\Delta = \delta^2 D(\Delta) S(\Delta;\psi^\Delta, u^\Delta, v^\Delta) + n_D^\Delta(\psi^\Delta)$$

$$(SDFEM3) \quad -B_v(\Delta;\psi^\Delta)v^\Delta = \delta^2 D(\Delta) S(\Delta;\psi^\Delta, u^\Delta, v^\Delta) + p_D^\Delta(\psi^\Delta)$$

The diagonal matrix $E(\psi^\Delta)$ has the entries e^{ψ_i} and B_u, B_v, S are related to B_n, B_p and R resp. by:

$$B_u(\Delta;\psi^\Delta) := \delta^2 B_n(\Delta;\psi^\Delta)E(\psi^\Delta),$$

$$B_v(\Delta;\psi^\Delta) := \delta^2 B_p(\Delta;\psi^\Delta)E^{-1}(\psi^\Delta). \tag{5.2.1}$$

$$S(\Delta;\psi^\Delta, u^\Delta, v^\Delta) := \frac{1}{\delta^2} R(\Delta;\delta^2 E(\psi^\Delta)u^\Delta, \delta^2 E^{-1}(\psi^\Delta)v^\Delta). \tag{5.2.2}$$

For most practical computations the formulation *(SDFE)* is preferable to *(SDFEM)*, since the latter requires explicit evaluation of the exponentials e^{ψ_i}, $e^{-\psi_i}$. This can lead to over- and underflow for high injection applications. The entries of the matrices B_n and B_p depend only on exponentials of potential differences at neighbouring nodes and can usually be accurately evaluated.
Finite difference schemes are – after suitable premultiplication of the difference equations – obtained by appropriately perturbing the entries of D. In the two-dimensional case the (ij)-diagonal entry of the modified matrix D is $((h_{i+1}+h_i)/2)\cdot((k_{j+1}+k_j)/2)$.

The Coefficient Matrices

The analysis of the properties of the coefficient matrices is based on the concepts of M-matrices and Stieltjes matrices (see, e.g., [5.32]). We call a real $n\times n$-matrix A an M-matrix, if its off-diagonal entries are nonpositive, A is nonsingular, and if all entries of the inverse matrix A^{-1} are nonnegative. A is called a Stieltjes matrix, if its off-diagonal entries are nonpositive and if A is symmetric and positive definite. Note that every Stieltjes matrix is an M-matrix (see [5.32]).
In the sequel we shall need a restriction on the interior angles of the finite elements.
We call a simplicial mesh Δ of acute type, if all interior angles of all triangles or, resp., all interior angles between faces of tetrahedrons are not larger than $\pi/2$.
Zlamal showed in [5.35], [5.36] that the quadratic forms a, $b_{u,\Delta}$ and $b_{v,\Delta}$ satisfy a maximum principle on X_Δ, if the mesh is of acute type and if $\mu_n > 0$, $\mu_p > 0$.

Finite difference analogues of the discrete maximum principle can be found in [5.3], [5.7].
In the following Theorem we collect important properties of the coefficient matrices.

Theorem 5.2.1: *Let $\psi^\Delta \in \mathbb{R}^M$ and, for $\Omega \subseteq \mathbb{R}^2$ or $\Omega \subseteq \mathbb{R}^3$, assume that the mesh Δ is of acute type. Also assume that the $(k-1)$-dimensional Lebesgue measure of $\partial\Omega_D$ is positive, that $\partial\Omega_D$ contains at least one node and that $\mu_n > 0$, $\mu_p > 0$. Then $B_n(\Delta; \psi^\Delta)$, $B_p(\Delta; \psi^\Delta)$ are M-matrices and $A(\Delta)$, $B_u(\Delta; \psi^\Delta)$, $B_v(\Delta; \psi^\Delta)$ are diagonally dominant Stieltjes matrices.*

The main ideas of the proof can be found in [5.35].
Clearly, Theorem 5.2.1 carries over to the coefficient matrices of the discussed finite difference schemes.

Assume now that the recombination-generation rate is of Shockley-Read-Hall and/or Auger type and that the mesh is of acute type. Then the proof of existence of discrete solutions proceeds along the lines of the proof of the solvability of the continuous device problem as presented in Section 3.2, i.e. it is based on decoupling *(SDFEM1)*, *(SDFEM2)*, *(SDFEM3)* and applying Schauder's fixed point theorem and discrete maximum principle estimates. The unique solvability of the decoupled problem follows from Theorem 5.2.1. The obtained discrete solution $(\psi^\Delta, u^\Delta, v^\Delta)$ satisfies the L^∞-estimates of Section 3.2.
Two different approaches for the actual computational solution of the discrete system *(SDFE)* are in use. The first approach is represented by approximately solving the fully coupled system using a Newton or accelerated Newton iteration and the second is based on Gummel-type blockiteration methods. Both approaches are descibed in Section 3.6 on the continuous level.
In view of Theorem 5.2.1 the assertions of Section 3.6 can easily be carried over to the discrete device problem.
Newton-type methods are discussed in [5.29], [5.33] and blockiteration techniques in [5.5], [5.20], [5.2], [5.9], [5.13]. A method for choosing initial guesses for nonlinear iterations based on arclength continuation (see [5.27]) was presented in [5.18].

5.3 Convergence of Discretisations

In this section we seek answers to the following questions: firstly, do the finite element and finite difference solutions converge (in suitable norms) to a solution of the 'continuous' problem when the mesh is appropriately refined and secondly, how should one construct a mesh with as few nodes as possible such that the error between the discrete solution and a 'continuous' solution is less than some prescribed error tolerance? By employing a 'decoupled convergence analysis' we shall positively answer the first question.
The most serious problem for the construction of 'efficient' meshes is the occurence of thin layer strips about *pn*-junctions, Schottky contacts and

semiconductor-oxide interfaces, within which the solutions ψ, n, p vary rapidly (see Chapter 4 for details). Therefore, on first glance, it seems that the mesh has to be sufficiently fine within the layers in order to obtain accurate discrete solutions. In one space-dimension the solutions can be accurately resolved within layers without employing excessive computer resources, for multi-dimensional problems, however, the costs may become very high. We shall demonstrate that rather coarse, even layer-ignoring meshes very often allow for qualitatively correct solutions. The main reason for this lies in the special form of the *SG*-discretisations, whose convergence properties are based on accurately resolving the current densities but not on the accurate resolution of the quantities ψ, n and p.

Decoupled Convergence Analysis

Firstly, we assume that the recombination-generation rate R vanishes identically:

$$R \equiv 0, \tag{5.3.1}$$

and that the mobilities only depend on the position vector x and satisfy:

$$0 < \underline{\mu}_n \leqq \mu_n(x) \leqq \bar{\mu}_n, \qquad 0 < \underline{\mu}_p \leqq \mu_p(x) \leqq \bar{\mu}_p. \tag{5.3.2}$$

Moreover, the assumptions of Theorem 5.2.1 are assumed to hold, in particular the finite element mesh Δ is – for multi-dimensional problems – of acute type in order to guarantee the existence of discrete solutions.
We shall now decouple the static device problem. For a given finite element solution $(\psi_\Delta, u_\Delta, v_\Delta) \in (X_\Delta)^3$ we denote by $\psi_{\Delta,1}$ the solution of the perturbed Poisson's equation:

$$\lambda^2 \Delta \psi_{\Delta,1} = \delta^2 e^{\psi_{\Delta,1}} u_\Delta - \delta^2 e^{-\psi_{\Delta,1}} v_\Delta - C(x), \qquad x \in \Omega \tag{5.3.3 (a)}$$

$$\psi_{\Delta,1}\big|_{\partial\Omega_D} = \psi_D\big|_{\partial\Omega_D}, \quad \frac{\partial \psi_{\Delta,1}}{\partial \nu}\bigg|_{\partial\Omega_N} = 0, \tag{5.3.3 (b)}$$

and by $u(\psi_\Delta)$, $v(\psi_\Delta)$ the exact solutions of:

$$\operatorname{div}\left(\mu_n e^{\psi_\Delta} \operatorname{grad} u(\psi_\Delta)\right) = 0, \qquad x \in \Omega \tag{5.3.4 (a)}$$

$$u(\psi_\Delta)\big|_{\partial\Omega_D} = u_D\big|_{\partial\Omega_D}, \quad \frac{\partial u(\psi_\Delta)}{\partial \nu}\bigg|_{\partial\Omega_N} = 0 \tag{5.3.4 (b)}$$

and, resp.:

$$\operatorname{div}\left(\mu_p e^{-\psi_\Delta} \operatorname{grad} v(\psi_\Delta)\right) = 0, \qquad x \in \Omega \tag{5.3.5 (a)}$$

$$v(\psi_\Delta)\big|_{\partial\Omega_D} = v_D\big|_{\partial\Omega_D}, \quad \frac{\partial v(\psi_\Delta)}{\partial \nu}\bigg|_{\partial\Omega_N} = 0. \tag{5.3.5 (b)}$$

If the Frechet derivative of the static device problem at the discrete solution $(\psi_\Delta, u_\Delta, v_\Delta)$ is boundedly invertible (when regarded as map between appropriate spaces), then we derive from Theorem 3.6.3 that an isolated solution (ψ, u, v) of the continuous problem and constants $D_1, \chi_1 > 0$ exist such that

$$\| \psi - \psi_\Delta \|_{\infty,\Omega} \leqq D_1(\| \psi_\Delta - \psi_{\Delta,1} \|_{\infty,\Omega} + \| u(\psi_\Delta) - u_\Delta \|_{\infty,\Omega} + \| v(\psi_\Delta) - v_\Delta \|_{\infty,\Omega}) \tag{5.3.6}$$

holds, if the right hand side of (5.3.6) is less than χ_1.
An H^1-analogon of (5.3.6) was proven in [5.21], Section 2.10:

$$\| \psi - \psi_\Delta \|^{\lambda}_{1,2,\Omega} \leqq D_2(\| \psi_\Delta - \psi_{\Delta,1} \|^{\lambda}_{1,2,\Omega} + \| u(\psi_\Delta) - u_\Delta \|_{2,\Omega} + \| v(\psi_\Delta) - v_\Delta \|_{2,\Omega}), \tag{5.3.7}$$

if the right hand side of (5.3.7) is less than some constant χ_2. We denoted

$$\| \psi \|^{\lambda}_{1,2,\Omega} := \left(\int_\Omega |\psi|^2 \, dx + \lambda^2 \int_\Omega |\operatorname{grad} \psi|^2 \, dx\right)^{\frac{1}{2}}. \tag{5.3.8}$$

The norms on the right and left hand sides of (5.3.6), (5.3.7) are chosen such that it can be shown – with some extra effort – that the constants D_1 and D_2 are independent of λ as $\lambda \to 0+$. This is particularly important in view of the singular perturbation character of the device problem. The constants D_1, D_2 only weakly depend on δ as $\delta \to 0+$ and they are independent of the discretisation scheme. However, D_2 and D_1 may increase as the externally applied potentials increase.
Also, estimates for the approximation error of u and v are obtained from Section 3.6, e.g.:

$$\| u - u_\Delta \|_{2,\Omega} \leqq \| u_\Delta - u(\psi_\Delta) \|_{2,\Omega} + D_3 \| \psi - \psi_\Delta \|^{0,\Omega} \tag{5.3.9 (a)}$$

$$\| u - u_\Delta \|_{1,2,\Omega} \leqq \| u_\Delta - u(\psi_\Delta) \|_{1,2,\Omega} + D_4 \| \psi - \psi_\Delta \|^{0,\Omega} \tag{5.3.9 (b)}$$

and analogously for v. In one-space dimension, i.e. $\Omega = (-1, 1)$, L^∞-estimates hold, too:

$$\| u - u_\Delta \|^{0,(-1,1)} \leqq \| u_\Delta - u(\psi_\Delta) \|^{0,(-1,1)} + D_5 \| \psi - \psi_\Delta \|^{0,(-1,1)} \tag{5.3.10}$$

$$\| v - v_\Delta \|^{0,(-1,1)} \leqq \| v_\Delta - v(\psi_\Delta) \|^{0,(-1,1)} + D_6 \| \psi - \psi_\Delta \|^{0,(-1,1)}. \tag{5.3.11}$$

The constants $D_3, \ldots, D_6$ may depend strongly depend on δ. Typically, they blow up as $\delta \to 0+$.
Estimates for the discretisation error of the carrier concentrations n and p are easily derived. Therefore, we write:

$$n - n_\Delta = \delta^2 u(e^{\psi} - e^{\psi_\Delta}) + \delta^2 e^{\psi_\Delta}(u - u_\Delta) + \delta^2 e^{\psi_\Delta} u_\Delta - n_\Delta.$$

The first two expressions on the right hand side can be estimated by using (5.3.6)–(5.3.11) and the third term by observing that n_Δ is the piecewise linear interpolate of $\delta^2 e^{\psi_\Delta} u_\Delta$ on the mesh Δ. Clearly, the third term vanishes at the nodes.
The estimates (5.3.6), (5.3.7) were only rigorously proven for isolated solutions of the static device problem assuming a vanishing recombination-

generation rate and field-independent mobilities. Also, the proof of (3.5.6) requires strong regularity assumptions on the boundary conditions (see Theorem 3.6.3). The isolatedness assumptions can be weakened by applying the implicit function Theorem in the proof of Theorem 3.6.3 differently, e.g. to a reparametrised version of the device problem in order to include snap-back phenomena (see [5.18]), and – with some extra effort – the Shockley-Read-Hall recombination-generation term can be included, too. In order to avoid unnecessary complications we shall henceforth employ the estimates (5.3.6), (5.3.7) also for the convergence analysis of discretisations of more complicated device models without sticking to the formal assumptions. Clearly the equations (5.3.4) (a), (5.3.5) (a) have to be modified when field-dependent mobilities and nonvanishing recombination-generation terms are admitted, e.g. $\mu_{n,p}$ has to be substituted by $\mu_{n,p}(x, \operatorname{grad} \psi_\Delta)$.

Classical Convergence Theory

For multi-dimensional problems we shall only admit partitions, which are of acute type in order to guarantee the existence of finite element solutions, and – in order to guarantee the uniformity of the basis B_Δ of X_Δ (see [5.30]) – we assume that all angles between edges of triangles or, resp., between faces of tetrahedrons are bounded below by some positive angle θ_0 uniformly for all considered partitions Δ.

The convergence analysis of the multi-dimensional finite element SG-discretisations (5.1.53), (5.1.54) is highly complicated and goes well beyond the scope of this book (see [5.23], [5.24]). However, the main ideas can be demonstrated by analysing a simplified SG-type finite element discretisation. Instead of numbering the vertices of the finite elements in dependence of 'central' nodes we choose a predefined numbering of the vertices of each finite element and obtain uniquely defined discrete current densities J_{n,G_T}, J_{p,G_T} (see (5.1.50) (b), (c)) on each element. We multiply these discrete current densities by the gradient of a test-function $\varphi_\Delta \in X_{\Delta,0}$, integrate, and come up with (new) approximations to b_u, b_v:

$$b^*_{u,\Delta}(\psi_\Delta; u_\Delta, \varphi_\Delta) := \sum_{j=1}^{N} \delta^2 \mu_{n,G_{T_j}} \int_{T_j} F_{T_j}(\psi_\Delta) \operatorname{grad} u_\Delta \cdot \operatorname{grad} \varphi_\Delta \, dx \qquad (5.3.12)\ (a)$$

$$b^*_{v,\Delta}(\psi_\Delta; v_\Delta, \varphi_\Delta) := \sum_{j=1}^{N} \delta^2 \mu_{p,G_{T_j}} \int_{T_j} H_{T_j}(\psi_\Delta) \operatorname{grad} v_\Delta \cdot \operatorname{grad} \varphi_\Delta \, dx. \qquad (5.3.12)\ (b)$$

The new finite element discretisations of the combined current relations-continuity equations are obtained by substituting $b_{u,\Delta}$, $b_{v,\Delta}$ by $b^*_{u,\Delta}$ and $b^*_{v,\Delta}$ resp. in (5.1.53), (5.1.54).

In the one-dimensional case $b_{u,\Delta} = b^*_{u,\Delta}$, $b_{v,\Delta} = b^*_{v,\Delta}$ holds and the new discretisations coincide with (5.1.53), (5.1.54). In the multi-dimensional cases the quadratic forms $b^*_{u,\Delta}$, $b^*_{v,\Delta}$ are not symmetric (with respect to u_Δ, φ_Δ and v_Δ, φ_Δ resp.) and, thus, do not give practically useful finite element discretisations. Since $b^*_{u,\Delta}$, $b^*_{v,\Delta}$ have the same 'flavour' as $b_{u,\Delta}$, $b_{v,\Delta}$ their analysis, however, very well highlights qualitative properties of the SG-schemes.

In this subsection we denote by ψ_Δ, u_Δ, v_Δ the solutions of the modified discretisation scheme.

Firstly we shall bound the error $\| u(\psi_\Delta) - u_\Delta \|_{2,\Omega}$ in terms of the diameter d of the partition Δ. Therefore we denote by $\tilde{u}_\Delta$ the solution of the Ritz-Galerkin approximation to (5.1.36) with $\psi = \psi_\Delta$:

$$-b_u(\psi_\Delta; \tilde{u}_\Delta, \varphi_\Delta) = (R, \varphi_\Delta)_{2,\Omega} \quad \text{for all } \varphi \in X_{\Delta,0} \tag{5.3.13 (a)}$$

$$\tilde{u}_\Delta|_{\partial\Omega_D} = (u_D)_\Delta|_{\partial\Omega}. \tag{5.3.13 (b)}$$

A simple computation gives:

$$b^*_{u,\Delta}(\psi_\Delta; \tilde{u}_\Delta - u_\Delta, \tilde{u}_\Delta - u_\Delta) = (b^*_{u,\Delta} - b_u)(\psi_\Delta; \tilde{u}_\Delta, \tilde{u}_\Delta - u_\Delta)$$
$$+ ((R, \tilde{u}_\Delta - u_\Delta)_{2,\Omega,\Delta} - (R, \tilde{u}_\Delta - u_\Delta)_{2,\Omega}). \tag{5.3.14 (a)}$$

We set $b_{u,\Delta} = b_u + \varepsilon$ and estimate the quadratic form ε:

$$|\varepsilon(\psi_\Delta; w_\Delta, \varphi_\Delta)| \leqq \delta^2 \sup_\Omega e^{\psi_D} \cdot \max_l \Big(\|I_{T_l}\| \, \|I_{T_l}^{-1}\| \max_{j=1,\dots,k+1} O(\|\psi_\Delta - \psi_\Delta(P_{l_j})\|_{\infty,T_l})$$
$$+ O(\|\mu_{n,G_{T_l}} - \mu_n\|_{\infty,T_l}) \Big) \|w_\Delta\|_{1,2,\Omega} \|\varphi_\Delta\|_{1,2,\Omega} \tag{5.3.14 (b)}$$

for all w_Δ, $\varphi_\Delta \in H^1(\Omega)$. If the mesh Δ is fine enough such that ψ_Δ and μ_n are resolved accurately, i.e. if

$$|\psi_\Delta(P_i) - \psi_\Delta(P_j)| \leqq \sigma_0; \; P_i, P_j \in T_l, \quad l = 1, \dots, N \tag{5.3.15}$$

$$|\mu_{n,G_{T_l}} - \mu_n(x)| \leqq \sigma_0, \quad l = 1, \dots, N \tag{5.3.16}$$

holds for some sufficiently small constant $\sigma_0 > 0$, then we obtain from (5.3.14):

$$b^*_{u,\Delta}(\psi_\Delta; \tilde{u}_\Delta - u_\Delta, \tilde{u}_\Delta - u_\Delta) \geqq D_7 \| \tilde{u}_\Delta - u_\Delta \|^2_{1,2,\Omega}, \tag{5.3.17}$$

where $D_7 > 0$ only depends on the smallest angle θ_0, on σ_0, on the geometry, on μ_n and on the externally applied potentials since $\delta^2 \sup_\Omega e^{\psi_D} \leqq \text{const.}\, e^{|U_{\max}|}$.

Note that $\max_l (\|I_{T_l}\| \, \|I_{T_l}^{-1}\|)$ is bounded independently of the mesh because of the imposed angle restrictions.

By using the well-known accuracy results for the numerical integration formulas:

$$\left| \int_T f(x)\,dx - f(x_{G_T})\mu_k(T) \right| \leqq C_1 (\operatorname{diam} T)^2 \mu_k(T) |f|^{2,T} \tag{5.3.18 (a)}$$

$$\left| \int_T f(x)\,dx - \frac{\mu_k(T)}{k+1} \sum_{j=1}^{k+1} f_j \right| \leqq C_2 (\operatorname{diam} T)^2 \mu_k(T) |f|^{2,T}, \tag{5.3.18 (b)}$$

and

$$\| F_T(\psi_\Delta) - e^{\psi_\Delta(x_{G_T})} I_k \| \leqq C_3 (\operatorname{diam} T)^2 |e^{\psi_\Delta}|^{2,T}, \tag{5.3.18 (c)}$$

we derive the following estimates for the right hand side of (5.3.14) (see, e.g., [5.6], [5.30] for details):

$$|(b^*_{u,\Delta}-b_u)(\psi_\Delta;\tilde{u}_\Delta,\tilde{u}_\Delta-u_\Delta)| \leqq D_8 \max_l (d_l^2|\mu_n|^{2,T_l}+d_l\|e^{\psi_\Delta}\|^{1,T_l}$$
$$+d_l^2|e^{\psi_\Delta}|^{2,T_l})\,\|\tilde{u}_\Delta-u_\Delta\|_{1,2,\Omega}, \qquad (5.3.19)\ (a)$$

$$|(R,\tilde{u}_\Delta-u_\Delta)_{2,\Omega}-(R,\tilde{u}_\Delta-u_\Delta)_{2,\Omega,\Delta}|$$
$$\leqq D_9 \max_l d_l^2\,\|R\|^{2,T_l}\,\|\tilde{u}_\Delta-u_\Delta\|_{1,2,\Omega}. \qquad (5.3.19)\ (b)$$

Combining (5.3.14), (5.3.17), (5.3.19) gives, assuming that ψ_Δ and μ_n are accurately resolved:

$$\|\tilde{u}_\Delta-u_\Delta\|_{1,2,\Omega} \leqq D_{10} \max_l (d_l^2(|\mu_n|^{2,T_l}+|e^{\psi_\Delta}|^{2,T_l}+\|R\|^{2,T_l})$$
$$+d_l\|e^{\psi_\Delta}\|^{1,T_l}). \qquad (5.3.20)$$

By applying the standard interpolation error estimate to $u(\psi_\Delta)-\tilde{u}_\Delta$, which takes into account the error caused by interpolating u_Δ by $(u_D)_\Delta$ (see [5.30]), and by using the coercivity of the quadratic form b_u we obtain:

$$\|u(\psi_\Delta)-u_\Delta\|_{1,2,\Omega} \leqq D_{11} \max_l (d_l(\mu_k(T_l))^{-\frac{1}{2}}|u(\psi_\Delta)|_{2,2,T_l}+\|e^{\psi_\Delta}\|^{1,T_l})$$
$$+d_l^2(|\mu_n|^{2,T_l}+|e^{\psi_\Delta}|^{2,T_l}+\|R\|^{2,T_l})). \qquad (5.3.21)$$

Here we ignored boundary singularities and assumed $u(\psi_\Delta)\in H^2(\Omega)$.
An analogous estimate is obtained for $v(\psi_\Delta)-v_\Delta$, if μ_p is resolved accurately by the mesh, only $v(\psi_\Delta)$, μ_p, $e^{-\psi_\Delta}$ have to be substituted for $u(\psi_\Delta)$, μ_n and e^{ψ_Δ} resp. in the right hand side of (5.3.21).
The error $\psi_\Delta-\psi_{\Delta,1}$ of the semilinear Poisson's equation can be estimated in a similar way. Again, it consists of the interpolation error and of the error caused by numerical integration of the products of the space-charge density and $H^1(\Omega)$-test-functions:

$$\|\psi_\Delta-\psi_{\Delta,1}\|^{\lambda}_{1,2,\Omega} \leqq D_{12} \max_l ((d_l^2+\lambda d_l)(\mu_k(T_l))^{-\frac{1}{2}}|\psi_{\Delta,1}|_{2,2,T_l}$$
$$+d_l\|\varrho_{\Delta,1}\|^{2,T_l})) \qquad (5.3.22)$$

assuming that $\psi_{\Delta,1}\in H^2(\Omega)$, $\varrho_{\Delta,1}\in C^2(\bar{T}_l)$, $l=1,\ldots,N$, where we denoted:

$$\varrho_{\Delta,1} = -(\delta^2 e^{\psi_{\Delta,1}}u_\Delta-\delta^2 e^{-\psi_{\Delta,1}}v_\Delta-C). \qquad (5.3.23)$$

We combine the estimates (5.3.21), (5.3.22) and use (5.3.7) to obtain an $H^1(\Omega)$-estimate for the discretisation error of the coupled device problem:

$$\|\psi-\psi_\Delta\|^{\lambda}_{1,2,\Omega} \leqq D \max_l ((d_l^2+\lambda d_l)\mu_k(T_l)^{-\frac{1}{2}}|\psi_{\Delta,1}|_{2,2,T_l}+d_l(\|\varrho_{\Delta,1}\|^{2,T_l}$$
$$+\mu_k(T_l)^{-\frac{1}{2}}|u(\psi_\Delta)|_{2,2,T_l}+\mu_k(T_l)^{-\frac{1}{2}}|v(\psi_\Delta)|_{2,2,T_l}$$
$$+\|e^{\psi_\Delta}\|^{1,T_l}+\|e^{-\psi_\Delta}\|^{1,T_l})+d_l^2(|\mu_n|^{2,T_l}$$
$$+|\mu_p|^{2,T_l}+|e^{\psi_\Delta}|^{2,T_l}+|e^{-\psi_\Delta}|^{2,T_l}+\|R\|^{2,T_l})). \qquad (5.3.24)$$

Note that the accurate resolution of the potential and the mobilities by the mesh was used to prove (5.3.24).
The estimate says, that under the above mentioned assumptions, the finite element discretisation is convergent of order one in the $\|\cdot\|^{\lambda}_{1,2,\Omega}$-norm, i.e.

$$\|\psi-\psi_\Delta\|^{\lambda}_{1,2,\Omega} = O(d) \tag{5.3.25}$$

with $d = \max_l d_l$. Clearly, it also converges of order one in the $H^1(\Omega)$-norm, then, however, the coefficient D is expected to blow up as $\lambda \to 0+$.
By using standard methods from finite element analysis, improved rates of convergence can be proven in the $L^2(\Omega)$-norm:

$$\|\psi-\psi_\Delta\|_{2,\Omega} = O(d^2), \qquad (5.3.26)\ (a)$$

and – by using (5.3.6) – in the $L^\infty(\Omega)$-norm, if difficulties steming from boundary singularities are ignored and if the mesh is refined such that $c \leqq d_l/d \leqq 1$ holds for some constant $c > 0$:

$$\|\psi-\psi_\Delta\|_{\infty,\Omega} = O\left(d^2 \ln\frac{1}{d}\right) \qquad (5.3.26)\ (b)$$

(see [5.6], [5.30]).
Error estimates for u, v, n and p follow from (5.3.9)–(5.3.12) by using (5.3.21), (5.3.26):

$$\begin{aligned} \|u-u_\Delta\|_{2,\Omega}+\|v-v_\Delta\|_{2,\Omega} &= O\left(d^2 \ln\frac{1}{d}\right), \\ \|u-u_\Delta\|_{1,2,\Omega}+\|v-v_\Delta\|_{1,2,\Omega} &= O(d) \end{aligned} \qquad (5.3.27)\ (a)$$

$$\|n-n_\Delta\|_{2,\Omega}+\|p-p_\Delta\|_{2,\Omega} = O\left(d^2 \ln\frac{1}{d}\right). \qquad (5.3.27)\ (b)$$

This convergence analysis implies that the finite element solution ψ_Δ approximates the 'continuous' potential ψ with an error $O(\chi)$, where χ is a prescribed error tolerance, if – for every finite element T_l – the expressions $(d_l^2+\lambda d_l)\mu_k(T_l)^{-\frac{1}{2}}|\psi_{\Delta,1}|_{2,2,T_l}, \ldots, d_l^2\|R\|^{2,T_l}$ are of the order of magnitude χ. Thus, the potential $\psi_{\Delta,1}$, the space charge density $\varrho_{\Delta,1}$, the mobilities μ_n, μ_p, the recombination-generation rate R and $u(\psi_\Delta)$, $v(\psi_\Delta)$ are required to vary moderately on each finite element. In practice this is only very difficult or even impossible to guarantee, since the singular perturbation character of the static device problem causes ψ and ϱ to vary rapidly within space charge-layers (see Chapter 4), thus requiring a very fine mesh there, which may easily lead to excessive computer storage and time requirements for realistic multi-dimensional applications.
Thus, the 'classical' convergence estimates (5.3.24)–(5.3.27) are – from a practical point of view – unsatisfactory and it is our next goal to improve them by showing that it is not necessary to accurately resolve all quantities in (5.3.24) in order to achieve reasonable numerical results.
The clou to this lies in the observation that the presented analysis did not at

all make use of the special form of the SG-discretisation. In fact, we interpreted the SG-schemes as perturbations of standard piecewise linear finite element discretisations of the combined current relations-continuity equations and obtained the same convergence results as for the standard discretisations. Thus, the above analysis only tells us that the SG-schemes converge at least as well as the standard discretisations. However, due to their special construction, they should perform better. Because of the exponential fitting employed to set up the SG-schemes, we expect them to give accurate approximations even if the potential and the carrier concentrations are not resolved well by the mesh.

Convergence Analysis of the SG-Schemes

In order to better understand the performance of the SG-schemes we now present a convergence analysis for the one-dimensional problem.
For simplicity we assume that R, μ_n, μ_p are smooth functions of the space variable x only, i.e. $R = R(x)$, $\mu_n = \mu_n(x)$, $\mu_p = \mu_p(x)$.
For a given piecewise linear function ψ_Δ, the combined current-relation continuity equation for electrons reads:

$$(\delta^2 \mu_n(x) e^{\psi_\Delta} u(\psi_\Delta)')' = R(x), \qquad -1 \leqq x \leqq 1 \tag{5.3.28 (a)}$$

$$u(\psi_\Delta)(-1) = u_D(-1), \qquad u(\psi_\Delta)(1) = u_D(1) \tag{5.3.28 (b)}$$

and its SG-discretisation is given by the one-dimensional version of (5.1.67). The 'continuous' problem as well as the 'discrete' problem can easily be solved exactly:

$$u(\psi_\Delta)(x) = u_D(-1) + \frac{u_D(1) - u_D(-1)}{\int\limits_{-1}^{1} \frac{e^{-\psi_\Delta}}{\mu_n} ds} \int\limits_{-1}^{x} \frac{e^{-\psi_\Delta}}{\mu_n} ds + F(x) - F(1) \frac{\int\limits_{-1}^{x} \frac{e^{-\psi_\Delta}}{\mu_n} ds}{\int\limits_{-1}^{1} \frac{e^{-\psi_\Delta}}{\mu_n} ds}, \tag{5.3.29 (a)}$$

where

$$F(x) := \int\limits_{-1}^{x} \frac{e^{-\psi_\Delta(s)}}{\mu_n(s)} \int\limits_{-1}^{s} S(\tau)\, d\tau\, ds, \qquad S(x) := \frac{1}{\delta^2} R(x) \tag{5.3.29 (b)}$$

and, resp.:

$$u_i = u_D(-1) + \frac{u_D(1) - u_D(-1)}{\sum\limits_{l=0}^{N-1} \frac{h_{l+1}}{\mu_{n,l+\frac{1}{2}} \sigma_{l+\frac{1}{2}}}} \sum_{l=0}^{i-1} \frac{h_{l+1}}{\mu_{n,l+\frac{1}{2}} \sigma_{l+\frac{1}{2}}}$$

$$+F_i - F_N \frac{\sum_{l=0}^{i-1} \frac{h_{l+1}}{\mu_{n,l+\frac{1}{2}} \sigma_{l+\frac{1}{2}}}}{\sum_{l=0}^{N-1} \frac{h_{l+1}}{\mu_{n,l+\frac{1}{2}} \sigma_{l+\frac{1}{2}}}}, \tag{5.3.30 (a)}$$

where $F_1 = 0$ and

$$F_i := \sum_{l=0}^{i-1} \frac{h_{l+1}}{\mu_{n,l+\frac{1}{2}} \sigma_{l+\frac{1}{2}}} \sum_{j=1}^{l} \frac{h_{j+1}+h_j}{2} S_j,$$

$$S_i = \frac{1}{\delta^2} R_i, \tag{5.3.30 (b)}$$

$$\sigma_{i+\frac{1}{2}} := e^{\psi_{i+1}} B(\psi_{i+1} - \psi_i). \tag{5.3.30 (c)}$$

In order to estimate the error $|u(\psi_\Delta)(x_i) - u_i|$ we only have to bound the differences

$$\xi_i := \int_{-1}^{x_i} \frac{e^{-\psi_\Delta(s)}}{\mu_n(s)} ds - \sum_{l=0}^{i-1} \frac{h_{l+1}}{\mu_{n,l+\frac{1}{2}} \sigma_{l+\frac{1}{2}}} \tag{5.3.31 (a)}$$

$$\eta_i := F(x_i) - F_i \tag{5.3.31 (b)}$$

in terms of the meshsizes. Since ψ_Δ is piecewise linear we have

$$\int_{x_l}^{x_{l+1}} e^{-\psi_\Delta(s)} ds = \frac{h_{l+1}}{\sigma_{l+\frac{1}{2}}}, \tag{5.3.32}$$

and therefore

$$\xi_i = \sum_{l=0}^{i-1} \int_{x_l}^{x_{l+1}} \left(\frac{1}{\mu_n(s)} - \frac{1}{\mu_{n,l+\frac{1}{2}}} \right) e^{-\psi_\Delta(s)} ds$$

follows. Obviously the estimate

$$|\xi_i| \leq E_1 \max_{l=1,\ldots,i} \left(h_l \left\| \frac{\mu_n'}{\mu_n^2} \right\|^{0,[x_{l-1},x_l]} \right) \tag{5.3.33 (a)}$$

holds independently of the variation of $e^{-\psi_\Delta}$ on each subinterval. By proceeding similarly we derive:

$$|\eta_i| \leq E_2 \max_{l=1,\ldots,i} \left(h_l \left(\left\| \frac{\mu_n'}{\mu_n} \right\|^{0,[x_{l-1},x_l]} + \left\| \frac{R}{\mu_n} \right\|^{1,[x_{l-1},x_l]} \right) \right) \tag{5.3.33 (b)}$$

and, by using $|u(\psi_\Delta)(x_i) - u_i| = O\left(\max_l (|\xi_l| + |\eta_l|)\right)$, we obtain

$$|u(\psi_\Delta)(x_i) - u_i|$$
$$\leq E_3 \max_l \left(h_l \left(\left\| \frac{\mu_n'}{\mu_n} \right\|^{0,[x_{l-1},x_l]} + \left\| \frac{R}{\mu_n} \right\|^{1,[x_{l-1},x_l]} \right) \right). \tag{5.3.34}$$

The constant E_3 – and therefore the error bound (5.3.34) – depends on the applied potentials but not on the differences $\psi_{l+1}-\psi_l$. We remark that, by allowing arbitrarily large values $\psi_{l+1}-\psi_l$, the order of convergence is reduced to 1. For constant μ_n and vanishing R, the discrete solutions u_i of the SG-scheme are the exact nodal values $u(\psi_\Delta)(x_i)$.
Clearly, second derivatives appear in the global error estimates for the piecewise linear functions ψ_Δ, u_Δ and v_Δ. Thus we obtain from (5.3.6):

$$\|\psi-\psi_\Delta\|^{0,[-1,1]} \leqq E_5\left(\max_l |\psi_l-\psi_{\Delta,1}(x_l)|+h_l\left(\left\|\frac{\mu_n'}{\underline{\mu}_n}\right\|^{0,[x_{l-1},x_l]}\right.\right.$$

$$+\left\|\frac{\mu_p'}{\underline{\mu}_p}\right\|^{0,[x_{l-1},x_l]}+\left.\left\|\frac{R}{\underline{\mu}}\right\|^{1,[x_{l-1},x_l]}\right)+h_l^2(\|\psi_{\Delta,1}''\|^{0,[x_{l-1},x_l]}$$

$$\left.+\|u(\psi_\Delta)''\|^{0,[x_{l-1},x_l]}+\|v(\psi_\Delta)''\|^{0,[x_{l-1},x_l]})\right), \tag{5.3.35}$$

where $\underline{\mu} = \min(\underline{\mu}_n, \underline{\mu}_p)$. The lower bounds $\underline{\mu}_n$, $\underline{\mu}_p$ of the mobilities were not incorporated into the constants $E_1, \ldots, E_5$ in order to emphasize the scaling invariance of the estimates.
The first term on the right hand side represents the error (at the nodes) of the discretisation of Poisson's equation. It will be analysed in more detail later on.
Fast varying quantities – like the potential and the space charge density – have to be accurately resolved in order to obtain convergence in the whole interval $[-1, 1]$. Thus the estimate (5.3.35) is – in flavour – very similar to the estimate (5.3.24) and it still does not demonstrate the 'power' of the SG-discretisation in a satisfactory way. Therefore, we have to look at estimates of the nodal value differences only.
Usually, it is possible to resolve the potential ψ accurately in outer regions, i.e. outside the layers within which the solutions ψ, n and p vary rapidly. As discussed in Chapter 4 the layer regions are usually thin in direction orthogonal to critical surfaces. Accurately speaking, the width of a *pn*-junction layer is $O(\lambda\,|\ln\lambda|)$, if λ is sufficiently small and if the junction is exponentially graded. Thus, assuming that the potential is approximated by the piecewise linear finite element solution ψ_Δ to accuracy χ_q in the L^q-norm *outside all layer regions,* which occur in the device under consideration, we obtain the estimate:

$$\|\psi-\psi_\Delta\|_{q,\Omega} \leqq \|\psi-\psi_\Delta\|_{q,\Omega-S}+\|\psi-\psi_\Delta\|_{q,S}$$

$$\leqq \chi_q+O\left((\lambda\,|\ln\lambda|)^{\frac{1}{q}}\right), \tag{5.3.36}$$

since $\mu_k(S) = O(\lambda\,|\ln\lambda|)$, where we denoted by S the union of all layer regions in the device. Then, in the k-dimensional case, the corresponding solutions u, v satisfy:

$$\|u(\psi)-u(\psi_\Delta)\|_{\infty,\Omega} \leqq K_1\|(e^{\psi}-e^{\psi_\Delta})|\operatorname{grad} u|\,\|_{r,\Omega}$$

$$\leqq K_2\|\psi-\psi_\Delta\|_{q,\Omega}\|\,|\operatorname{grad} u|\,\|_{\frac{rq}{q-r}}, \tag{5.3.37}$$

$$\|v(\psi)-v(\psi_\Delta)\|_{\infty,\Omega} \leqq K_3\|(e^{-\psi}-e^{-\psi_\Delta})|\operatorname{grad} v|\,\|_{q,\Omega}$$
$$\leqq K_4\|\psi-\psi_\Delta\|_{q,\Omega}\,\|\,|\operatorname{grad} v|\,\|_{\frac{rq}{q-r}},$$

$$\text{for} \quad q > r > k \tag{5.3.38}$$

(see [5.12], Section 8.6), assuming that $u, v \in W^{1,\frac{rq}{q-r}}(\Omega)$. We conclude from (5.3.36):

$$\|u(\psi)-u(\psi_\Delta)\|_{\infty,\Omega} = O(\chi_q)+O\big((\lambda|\ln\lambda|)^{\frac{1}{q}}\big), \tag{5.3.39}$$

$$\|v(\psi)-v(\psi_\Delta)\|_{\infty,\Omega} = O(\chi_q)+O\big((\lambda|\ln\lambda|)^{\frac{1}{q}}\big), \tag{5.3.40}$$

if $q > k$ and $u, v \in W^{1,s}(\Omega)$ for some s sufficiently large.
Assume now that the grid Δ is chosen such that

$$\max_l h_l\left(\left\|\frac{\mu_n'}{\mu_n}\right\|^{0,[x_{l-1},x_l]}+\left\|\frac{\mu_p'}{\mu_p}\right\|^{0,[x_{l-1},x_l]}+\left\|\frac{R}{\mu}\right\|^{1,[x_{l-1},x_l]}\right) \leqq \gamma \tag{5.3.41}$$

holds for some prescribed error tolerance γ. Then, by combining (5.3.39), (5.3.40) and (5.3.34), (5.3.35) we obtain error estimates for the nodal values in the one-dimensional case:

$$\max_l |u(x_l)-u_l| = O(\chi_q)+O(\gamma)+O\big((\lambda|\ln\lambda|)^{\frac{1}{q}}\big) \tag{5.3.42 (a)}$$

$$\max_l |v(x_l)-v_l| = O(\chi_q)+O(\gamma)+O\big((\lambda|\ln\lambda|)^{\frac{1}{q}}\big). \tag{5.3.42 (b)}$$

The error bounds are independent of the resolution of the potential inside the layers.
In order to derive estimates for the approximation error of the carrier concentrations, we additionally assume that an estimate of the type (5.3.36) also holds in the discrete L^q-norm:

$$\|\psi-\psi_\Delta\|_{q,\Omega,\Delta} = O(\chi_q)+O\big((\lambda|\ln\lambda|)^{\frac{1}{q}}\big), \tag{5.3.43}$$

where we set

$$\|f\|_{q,\Omega,\Delta} = \left(\sum_{l=1}^{N}\mu_k(T_l)\,|f(P_{l_1})|^q\right)^{\frac{1}{q}}. \tag{5.3.44}$$

Here P_{l_1} denotes a node of the element T_l.
We easily obtain for $k = 1$, $\Omega = (-1, 1)$:

$$\|n-n_\Delta\|_{q,\Omega,\Delta} = O(\chi_q)+O(\gamma)+O\big((\lambda|\ln\lambda|)^{\frac{1}{q}}\big) \tag{5.3.45 (a)}$$

$$\|p-p_\Delta\|_{q,\Omega,\Delta} = O(\chi_q)+O(\gamma)+O\big((\lambda|\ln\lambda|)^{\frac{1}{q}}\big). \tag{5.3.45 (b)}$$

If ψ_Δ approximates ψ uniformly with the accuracy χ_∞ outside the layers, then the L^∞-estimates hold:

$$\max_{x_l \in \Omega - S} |n(x_l) - n_l| = O(\chi_\infty) + O(\gamma) + O\left((\lambda |\ln \lambda|)^{\frac{1}{q}}\right) \tag{5.3.46 (a)}$$

$$\max_{x_l \in \Omega - S} |p(x_l) - p_l| = O(\chi_\infty) + O(\gamma) + O\left((\lambda |\ln \lambda|)^{\frac{1}{q}}\right) \tag{5.3.46 (b)}$$

for every $q > 1$.
Estimates for the nodal errors of u and v for two-dimensional SG-schemes were obtained by M. Mock [5.23] by interpreting the schemes as mixed finite element methods (see [5.30]). He proved estimates of the form

$$\max_l |u(\psi_\Delta)(P_l) - u_l| = O(d), \tag{5.3.47 (a)}$$

$$\max_l |v(\psi_\Delta)(P_l) - v_l| = O(d) \tag{5.3.47 (b)}$$

under the assumption that the mesh Δ is sufficiently fine to resolve the current densities accurately, too.
This mesh restriction comes naturally, since the derivation of the SG-schemes is based on approximating the current densities by a constant on each finite element. Note that it may require a relatively fine mesh inside layers for multidimensional problems because – as shown in Chapter 4 – the current density components, which are tangential to the junctions, may vary rapidly inside the layer regions.
The proof of (5.3.47) is rather complicated and goes well beyond the scope of this book. We refer the reader to [5.23].
The estimates (5.3.42), (5.3.45), (5.3.46) carry over to the two-dimensional case when γ is substituted by d and if (5.3.36), (5.3.43) hold for some $q > k$. Apparently, the accuracy of the discretisation of the coupled static device problem hinges very much on how well the potential ψ can be approximated by the 'discrete' solution ψ_Δ outside layer regions.

Convergence Analysis for Poisson's Equation

In this paragraph we shall investigate qualitative and some quantitative features of the discretisation (5.1.16) of Poisson's equation. For the sake of simplicity we shall not try to be completely rigorous from the point of view of the convergence analysis of the coupled device problem since we shall neglect the errors caused by the coupling of Poisson's equation with the combined current relations-continuity equations. We assume that u and v are available exactly and just analyse the error caused by discretising the 'exact' Poisson's equation in dependence on the mesh. Particular emphasis will be given to the impact of the singular perturbation character of Poisson's equation on the construction of meshes.
The 'continuous' problem reads:

$$\lambda^2 \Delta \psi = \delta^2 e^{\psi} u(x) - \delta^2 e^{-\psi} v(x) - C(x), \qquad x \in \Omega \tag{5.3.48 (a)}$$

$$\psi\big|_{\partial\Omega_D} = \psi_D\big|_{\partial\Omega_D}, \qquad \left.\frac{\partial \psi}{\partial \nu}\right|_{\partial\Omega_N} = 0, \tag{5.3.48 (b)}$$

where $0 < \lambda \ll 1$. We only consider the finite difference approximation of the two-dimensional problem:

$$\lambda^2(\Delta_2\psi)_{ij} = \delta^2 e^{\psi_{ij}}u_{ij} - \delta^2 e^{-\psi_{ij}}v_{ij} - C_{ij}$$

$$\text{for all } (x_i, y_j) \in \Omega \cup \partial\Omega_N, \tag{5.3.49 (a)}$$

$$\psi_{ij} = \psi_D(x_i, y_j), \quad (x_i, y_j) \in \partial\Omega_D, \tag{5.3.49 (b)}$$

$$\begin{pmatrix} \dfrac{\psi_{i+1,j} - \psi_{i-1,j}}{2h_i} \\ \dfrac{\psi_{i,j+1} - \psi_{i,j-1}}{2k_j} \end{pmatrix} \cdot \nu = \begin{pmatrix} 0 \\ 0 \end{pmatrix}, \quad (x_i, y_j) \in \partial\Omega_N, \tag{5.3.49 (c)}$$

where we set $u_{ij} = u(x_i, y_j)$, $v_{ij} = v(x_i, y_j)$, $C_{ij} = C(x_i, y_j)$. Ω is a rectangle in $\mathbb{R}^2$, i and j run from 0 to N_x and 0 to N_y resp. and $h_{N_x+1} := h_{N_x}$, $h_0 := h_1$, $k_{N_y+1} := k_{N_y}$, $k_0 := k_1$. The 'discrete' Laplacian $(\Delta_2\psi)_{ij}$ is defined in (5.1.20). At first we compute the local discretisation error l_{ij} of (5.3.49) at a solution ψ of (5.3.48), defined by the residue of (5.3.49) at (x_i, y_j) when the 'continuous' solution ψ is inserted for the discrete solution. Taylor's formula gives at $(x_i, y_j) \in \Omega \cup \partial\Omega_N$ (after properly extending ψ 'beyond' $\partial\Omega_N$ in order to allow for the mirror-imaging):

$$\begin{aligned} |l_{ij}| \leqq \frac{\lambda^2}{3}\Bigg(& h_i \max_{x \in [x_{i-1}, x_i]} \left| \frac{\partial^3\psi}{\partial x^3}(x, y_j) \right| + h_{i+1} \max_{x \in [x_i, x_{i+1}]} \left| \frac{\partial^3\psi}{\partial x^3}(x, y_j) \right| \\ & + k_j \max_{y \in [y_{j-1}, y_j]} \left| \frac{\partial^3\psi}{\partial y^3}(x_i, y) \right| + k_{j+1} \max_{y \in [y_j, y_{j+1}]} \left| \frac{\partial^3\psi}{\partial y^3}(x_i, y) \right| \Bigg) \end{aligned}$$

$$\text{for } (x_i, y_j) \in \Omega \cup \partial\Omega_N. \tag{5.3.50}$$

For uniform or almost uniform grids defined by:

$$h_{i+1} = h_i(1 + O(h_i)), \quad k_{j+1} = k_j(1 + O(k_j)), \tag{5.3.51}$$

the estimate (5.3.50) can be improved such that $|l_{ij}|$ is $O(h_i^2 + k_j^2)$.
The local discretisation error vanishes at Dirichlet boundary points. At Neumann boundary points we have an additional local error contribution:

$$\begin{aligned} |l_{ij}^N| \leqq \frac{1}{6}\Bigg(& h_i^2 \max_{x \in [x_{i-1}, x_{i+1}]} \left| \frac{\partial^3\psi}{\partial x^3}(x, y_j) \right| \nu^x \\ & + k_j^2 \max_{y \in [y_{j-1}, y_{j+1}]} \left| \frac{\partial^3\psi}{\partial y^3}(x_i, y) \right| \nu^y \Bigg) \quad \text{for } (x_i, y_j) \in \partial\Omega_N, \end{aligned} \tag{5.3.52}$$

where we set $\nu = (\nu^x, \nu^y)'$.
By using Keller's stability-consistency theory [5.15] we conclude that the global error satisfies

$$\max_{i,j} |\psi(x_i, y_j) - \psi_{ij}| \leqq M \max_{i,j} |l_{ij}|, \tag{5.3.53}$$

if the right hand side is sufficiently small. The constant M is independent of the mesh and of λ, it depends on the boundary datum ψ_D, i.e. on the externally applied potentials, and weakly on δ. Typically $M = O(|\ln \delta|)$. For details we refer the interested reader to [5.17].

Let $0 < \kappa \ll 1$ be a given error tolerance. We now choose the mesh-sizes h_i such that $|l_{ij}| = O(\kappa)$, $|l_{ij}^N| = O(\kappa)$ at every grid point, i.e. by using (5.3.50), (5.3.52) such that

$$\lambda^2 \max_{x \in [x_{i-1}, x_i]} \left| \frac{\partial^3 \psi}{\partial x^3}(x, y_j) \right| h_i = O(\kappa), \tag{5.3.54 (a)}$$

$$\lambda^2 \max_{y \in [y_{i-1}, y_i]} \left| \frac{\partial^3 \psi}{\partial y^3}(x_i, y) \right| k_j = O(\kappa) \tag{5.3.54 (b)}$$

for $(x_i, y_j) \in \Omega \cup \partial\Omega_N$.

and

$$h_i^2 \max_{x \in [x_{i-1}, x_{i+1}]} \left| \frac{\partial^3 \psi}{\partial x^3}(x, y_j) \right| \nu^x + k_j^2 \max_{y \in [y_{j-1}, y_{j+1}]} \left| \frac{\partial^3 \psi}{\partial y^3}(x_i, y) \right| \nu^y = O(\kappa)$$

$$\text{for } (x_i, y_j) \in \partial\Omega_N. \tag{5.3.55}$$

Then we obtain from (5.3.53):

$$\max_{i,j} |\psi(x_i, y_j) - \psi_{ij}| = O(\kappa). \tag{5.3.56}$$

This method of constructing grids, which generate a global error of a prescribed order of magnitude by only controlling the local discretisation error, is called local error equidistribution.

In order to understand the drawback of equidistribution we consider the *pn*-junction diode depicted in Fig. 5.3.1. The *pn*-junction Γ is represented by the line segment $y = x$.

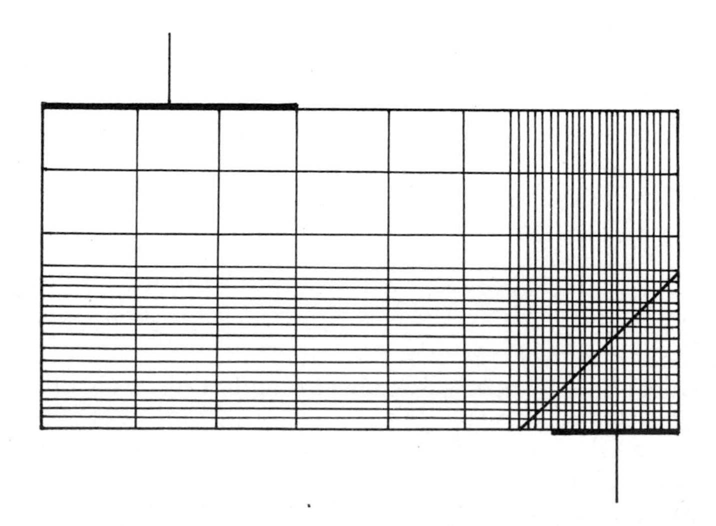

Fig. 5.3.1 *PN*-junction diode and equidistributing grid

From the singular perturbation analysis of Chapter 4 we derive the asymptotic expansion of the potential:

$$\psi(x, y, \lambda) = \bar{\psi}(x, y) + \hat{\psi}\left(\frac{w(x, y)}{\lambda}, s(x, y)\right) + O(\lambda), \tag{5.3.57 (a)}$$

$$\left|\frac{\partial^i \hat{\psi}}{\partial \omega^i}(\omega, s)\right| \leqq F_i \exp(-a|\omega|), \quad a > 0 \tag{5.3.57 (b)}$$

where

$$w(x, y) = \begin{cases} x - y, & x > y \\ y - x, & x \leqq y \end{cases}$$

denotes the 'orientiented' distance of (x, y) to Γ and $s(x, y)$ the point on Γ closest to (x, y). Differentiating (5.3.57) (a) gives:

$$\frac{\partial^3 \psi}{\partial x^3} = \frac{1}{\lambda^3}\frac{\partial^3 \hat{\psi}}{\partial \omega^3}\left(\frac{w(x, y)}{\lambda}, s(x, y)\right) w_x + O(\lambda^{-2}),$$

$$\frac{\partial^3 \psi}{\partial y^3} = \frac{1}{\lambda^3}\frac{\partial^3 \hat{\psi}}{\partial \omega^3}\left(\frac{w(x, y)}{\lambda}, s(x, y)\right) w_y + O(\lambda^{-2}) \tag{5.3.58}$$

and we obtain the equidistributing grid from (5.3.54):

$$h_i = O\left(\min\left(\kappa, \lambda\kappa \exp\left(-\frac{a|x_i - y_j|}{\lambda}\right)\right)\right),$$

$$k_j = O\left(\min\left(\kappa, \lambda\kappa \exp\left(-\frac{a|x_i - y_j|}{\lambda}\right)\right)\right). \tag{5.3.59}$$

The number of gridpoints inside the layer is at least:

$$N_{\text{layer}} = O\left(\frac{1}{\lambda\kappa^2}\right). \tag{5.3.60}$$

Since the x- and y-derivatives of ψ are large inside the layer strip, the meshsizes in x- and y-direction depend on λ and the number of gridpoints increases for constant error tolerance κ as λ decreases. Even when gridlines are allowed to terminate outside the layer, this leads to storage requirements, which cannot or can only hardly be met.
To see this, choose $\kappa = 10^{-2}$ and consider a device for which $\lambda \approx 10^{-3}$. Then $N_{\text{layer}} = O(10^7)$ and, taking into account the continuity equations, a linear system of dimension, which is of the order of magnitude 3×10^7, has to be solved in every Newton step.
Practically useful grids can usually only be obtained by local error equidistribution, if each device junction is parallel to one of the coordinate axes, i.e. in locally one-dimensional situations (see [5.16], [5.17]).
We shall now analyse how the convergence behaviour of discrete solutions is influenced when equidistribution is relaxed. Therefore, we consider a grid with

$$\underline{h} := \min_i h_i \geqq \beta\lambda, \quad \underline{k} := \min_j k_j \geqq \beta\lambda, \quad \beta > 0. \tag{5.3.61}$$

Then

$$\lambda^2 |(\Delta_2 \psi)_{ij}| \leqq \frac{8}{\beta^2} \max_{i,j} |\psi_{ij}| \tag{5.3.62}$$

holds and, if β is sufficiently large, a simple perturbation argument applied to (5.1.49) gives:

$$\psi_{ij} = \bar{\psi}(x_i, y_j) + O\left(\frac{\lambda^2}{\underline{h}^2} + \frac{\lambda^2}{\underline{k}^2}\right), \quad (x_i, y_j) \in \Omega. \tag{5.3.63}$$

Thus, if the grid is sufficiently coarse, the solution of the singularly perturbed scheme (5.3.49) approximates the reduced solution $\bar{\psi}$ accurately. Clearly, accuracy within layers cannot be expected.

The analysis of the singularly perturbed 3, 5 or 7-point schemes is extremly complicated when the grid is neither sufficiently fine inside the layer to guarantee equidistribution nor sufficiently coarse such that all meshsizes are much larger than the singular perturbation parameter. A convergence theory for general grids is not available. In order to qualitatively understand the behaviour of the discrete solutions in dependence on the grid, we investigate a one-dimensional linearised version of (5.3.48):

$$\lambda^2 \varphi'' = \alpha\varphi + C(x), \quad -1 \leqq x \leqq 1,$$

$$C(x) = \begin{cases} -1, & -1 \leqq x \leqq 0 \\ 1, & 0 < x \leqq 1 \end{cases}, \quad \alpha > 0 \tag{5.3.64 (a)}$$

$$\varphi(-1) = \frac{1}{\alpha}, \quad \varphi(1) = -\frac{1}{\alpha}. \tag{5.3.64 (b)}$$

The boundary data are chosen such that no layers occur there, i.e. $\varphi(-1) = \bar{\varphi}(-1)$, $\varphi(1) = \bar{\varphi}(1)$, where $\bar{\varphi} = C(x)/\alpha$ is the reduced solution. We assume that (5.3.64) is discretised by the 3-point scheme on a uniform grid with meshsize h and set

$$\frac{\lambda}{h} = \varrho. \tag{5.3.65}$$

Then the discretisation scheme reads:

$$\varrho^2(\varphi_{i+1} - 2\varphi_i + \varphi_{i-1}) = \alpha\varphi_i + C(x_i), \quad i = 1, \ldots, N-1 \tag{5.3.66 (a)}$$

$$\varphi_0 = \bar{\varphi}(-1), \quad \varphi_N = \bar{\varphi}(1). \tag{5.3.66 (b)}$$

We regard (5.3.66) (a) as two-step-recursion, set up the characteristic equation:

$$\varrho^2(r^2 - 2r + 1) - \alpha r = 0, \tag{5.3.67}$$

and compute the roots:

$$r_+(\varrho) = \frac{2\varrho^2 + \alpha + \sqrt{4\alpha\varrho^2 + \alpha^2}}{2\varrho^2}, \tag{5.3.68 (a)}$$

$$r_-(\varrho) = \frac{2\varrho^2 + \alpha - \sqrt{4\alpha\varrho^2 + \alpha^2}}{2\varrho^2}. \tag{5.3.68 (b)}$$

An easy calculation shows that $r_+ > 1$, $0 < r_- < 1$ for all $\varrho > 0$ and

$$\begin{aligned} r_+(0+) &= \infty, \quad r_+(\varrho) \downarrow 1 \quad \text{as} \quad \varrho \to \infty, \\ r_-(0+) &= 0, \quad r_-(\varrho) \uparrow 1 \quad \text{as} \quad \varrho \to \infty. \end{aligned} \tag{5.3.69}$$

We assume now that $N = 2L$ and that the gridpoint x_L coincides with the junction $x = 0$.
The solution of (5.3.66) is obtained explicitely by a simple computation. It is of the following form:

$$\varphi_i = \bar{\varphi}(x_i) + \left\{ \begin{array}{ll} \dfrac{2}{(r_+ - 1)\sqrt{4\alpha\varrho^2 + \alpha^2}} r_+^{i-L} & i \leqq L \\ \dfrac{2}{(1 - r_-)\sqrt{4\alpha\varrho^2 + \alpha^2}} r_-^{L-i}, & i > L \end{array} \right\} + O(r_+^{-L}) + O(r_-^L). \tag{5.3.70}$$

The term in parentheses decays to zero exponentially away from the junction, it is the 'discrete' junction layer term. Thus the structure of the 'discrete' solution φ_i is analogous to the structure of the 'continuous' solution $\varphi(x, \lambda)$ independently of the ratio ϱ of the perturbation parameter and the mesh size. The 'discrete' reduced solution approximates the 'continuous' reduced solution well, however, the 'discrete' internal layer term only approximates the 'continuous' internal layer term well (at the grid points), if h/λ is small. Therefore, the three-point-scheme is not uniformly (in λ) convergent, even the choice $h \approx \lambda$ creates a large global error within the layer. Outside the layer the global error is small independently of ϱ and therefore the 'discrete' solutions are – from a structural point of view – acceptable approximations for all $\varrho > 0$.
The presented discussion of the standard discretisation of Poisson's equation implies that even on a reasonably coarse grid the potential can be approximated accurately outside layer regions.

Current Density and Outflow Current Approximations

A great deal of the important information on the performance of devices can be extracted from the relationship of input voltages and outflow currents. Therefore, the computation of device currents is a crucial issue for virtually every device simulation code and the accuracy of the computed currents to a large extent determines the quality of the code.
Let $C_l \subseteq \partial\Omega_D$ denote a contact of the device under consideration. As discussed in Section 3.5 the outflow current at C_l can be written as

$$J_l = \int_\Omega J \cdot \operatorname{grad} \gamma_l \, dx, \tag{5.3.71}$$

where γ_l is an arbitrary $W^{1,\infty}(\Omega)$-function satisfying $\gamma_l = 1$ at C_l and $\gamma_l = 0$ at all other semiconductor contacts C_i. We use (5.1.30), (5.1.31) and (5.1.38), (5.1.39) in order to rewrite J_l as

$$J_l = b_n(\psi; n, \gamma_l) - b_p(\psi; p, \gamma_l), \tag{5.3.72 (a)}$$

$$J_l = b_u(\psi; u, \gamma_l) - b_v(\psi; v, \gamma_l). \tag{5.3.72 (b)}$$

We denote the piecewise linear interpolate of γ_l by $(\gamma_l)_\Delta$ and approximate the bilinear forms by their discrete counterparts in order to obtain an approximation $J_{l,\Delta}$ for J_l:

$$J_{l,\Delta} = b_{n,\Delta}(\psi_\Delta; n_\Delta, (\gamma_l)_\Delta) - b_{p,\Delta}(\psi_\Delta; p_\Delta, (\gamma_l)_\Delta), \tag{5.3.73}$$

or, expressed in the u, v-variables:

$$J_{l,\Delta} = b_{u,\Delta}(\psi_\Delta; u_\Delta, (\gamma_l)_\Delta) - b_{v,\Delta}(\psi_\Delta; v_\Delta, (\gamma_l)_\Delta). \tag{5.3.74}$$

When the finite element solutions ψ_Δ, n_Δ and p_Δ are known, then the outflow current approximations are computed as postprocessing by an additional evaluation of the 'discrete' forms $b_{n,\Delta}$ and $b_{p,\Delta}$. An arclength continuation algorithm for the computation of voltage-current characteristics, which allows the computation beyond turning points, was proposed in [5.18].
We assume now that the mesh is sufficiently fine such that all occuring fast varying quantities are accurately resolved and take the estimates (5.3.25)–(5.3.27) as starting point for the convergence analysis. We define the discrete current densities elementwise by

$$J_{n,\Delta}\big|_{T_l} := J_{n,G_{T_l}}, \qquad J_{p,\Delta}\big|_{T_l} := J_{p,G_{T_l}}, \tag{5.3.75}$$

where $J_{n,G_{T_l}}$, and $J_{p,G_{T_l}}$ are given by (5.1.56) and (5.1.57) resp. (assuming an arbitrary numbering of the nodes in T_l).A simple computation gives the $L^2(\Omega)$-estimates:

$$\|J_n - J_{n,\Delta}\|_{2,\Omega} = O\left(d \ln \frac{1}{d}\right) \tag{5.3.76 (a)}$$

$$\|J_p - J_{p,\Delta}\|_{2,\Omega} = O\left(d \ln \frac{1}{d}\right). \tag{5.3.76 (b)}$$

By using standard estimates an improved rate of convergence can be shown for the outflow currents:

$$|J_l - J_{l,\Delta}| = O\left(d^2 \ln \frac{1}{d}\right). \tag{5.3.77}$$

Again, the estimates (5.3.76), (5.3.77) are unsatisfactory since they require the accurate resolution of fast varying quantities by the mesh. In the one dimensional case it easily can be shown by using (5.3.29), (5.3.30) that is suffices to accurately resolve ψ outside the layers and μ_n, μ_p, R in Ω in order to obtain estimates for J_n, J_p (at the grid points) and for the total current J:

$$\max_l \left(|J_{n,l+\frac{1}{2}} - J_n(x_l)| + |J_{p,l+\frac{1}{2}} - J_p(x_l)| + |J - J_\Delta|\right)$$

$$= O(\chi_q) + O(\gamma) + O((\lambda|\ln \lambda|)^{\frac{1}{q}}), \tag{5.3.78}$$

if the grid is such that (5.3.41), (5.3.36) hold. In the multi-dimensional case the current densities J_n and J_p have to be accurately resolved, too (see [5.23], [5.24]).

5.4 Extensions and Conclusions

In this section we shall extend the discretisation schemes to allow for devices with semiconductor-oxide interfaces and Schottky contacts. We shall also discuss in detail the implications of the convergence analysis to the construction of 'efficient' meshes.

Semiconductor-Oxide Interfaces

Let Φ be an oxide-domain adjacent to Ω and let $\partial\Omega_I$ denote the interface. Then Laplace's equation

$$\Delta\psi = 0, \qquad x \in \Phi \tag{5.4.1}$$

and the interface conditions

$$[\psi]_{\partial\Omega_I} = \left[\varepsilon \frac{\partial\psi}{\partial\nu}\right]_{\partial\Omega_I} = 0, \qquad \varepsilon(x) := \begin{cases} \varepsilon_s, & x \in \Omega \\ \varepsilon_o, & x \in \Phi \end{cases} \tag{5.4.2}$$

hold. Also ψ satisfies homogeneous Neumann boundary conditions at the oxide boundary part $\partial\Phi_N$ and a Dirichlet boundary condition at the gate contact $\partial\Phi_D$ (see Section 2.3).
The weak formulation of Poisson's equation in the combined semiconductor-oxide domain $\Lambda := \Omega \cup \partial\Omega_I \cup \Phi$ then reads

$$-\lambda^2 a(\psi, \varphi) - \lambda^2 \frac{\varepsilon_o}{\varepsilon_s} \int_\Phi \operatorname{grad} \psi \cdot \operatorname{grad} \varphi \, dx = (n - p - C, \varphi)_{2,\Omega}$$

$$\text{for all } \varphi \in H_0^1(\Lambda \cup \partial\Lambda_D), \tag{5.4.3 (a)}$$

$$\psi - \psi_D \in H_0^1(\Lambda \cup \partial\Lambda_D), \tag{5.4.3 (b)}$$

where $\partial\Lambda_D = \partial\Omega_D \cup \partial\Phi_D$. We assume that $\partial\Phi$ is a union of finitely many line or, resp., plane segments and cover Φ by a simplicial mesh. We denote by Y_Δ the set of piecewise linear functions defined on Λ and by $Y_{\Delta,0}$ the subspace of Y_Δ consisting of all those functions, which vanish on $\partial\Lambda_D$. Then the piecewise linear finite element solution $\psi_\Delta \in Y_\Delta$ satisfies:

$$-\lambda^2 a(\psi_\Delta, \varphi_\Delta) - \lambda^2 \frac{\varepsilon_o}{\varepsilon_S} \int_\Phi \operatorname{grad} \psi_\Delta \cdot \operatorname{grad} \varphi_\Delta \, dx = (n_\Delta - p_\Delta - C, \varphi_\Delta)_{2,\Omega,\Delta}$$

for all $\varphi_\Delta \in Y_{\Delta,0}$, (5.4.4) (a)

$$\psi_\Delta|_{\partial A_D} = (\psi_D)_\Delta|_{\partial A_D}. \tag{5.4.4) (b}$$

Zero-current flow into the oxide is assumed:

$$J_n \cdot \nu|_{\partial\Omega_I} = J_p \cdot \nu|_{\partial\Omega_I} = 0. \tag{5.4.5}$$

Thus the finite element discretisations (5.1.61), (5.1.62) for the combined current relations-continuity equations remain valid when the interface $\partial\Omega_I$ is 'treated as an insulating semiconductor boundary segment'.
Finite difference discretisations of (5.4.1), (5.4.2) are obtained in a straightforward way.

Schottky Contacts

We now assume that the semiconductor device under consideration has Schottky contacts, whose union we denote by $\partial\Omega_S \subseteq \partial\Omega_D$. The following boundary conditions hold:

$$\psi|_{\partial\Omega_S} = \psi_D|_{\partial\Omega_S} \tag{5.4.6) (a}$$

$$J_n \cdot \nu|_{\partial\mathring{\Omega}_S} = -v_n(n - n_D)|_{\partial\mathring{\Omega}_S}, \qquad J_p \cdot \nu|_{\partial\mathring{\Omega}_S} = v_p(p - p_D)|_{\partial\mathring{\Omega}_S}, \tag{5.4.6) (b}$$

with v_n, v_p, $n_D|_{\partial\Omega_S}$, $p_D|_{\partial\Omega_S} > 0$ (see Section 2.3 for more information on the data).
Clearly, the finite element formulation (5.1.16) of Poisson's equation remains unchanged since $\partial\Omega_S$ is incorporated into $\partial\Omega_D$; the discretisations of the combined current relations-continuity equations, however, have to be modified. The weak formulation of (5.1.26), (5.4.6) (b) reads:

$$-b_n(\psi; n, \varphi) - \int_{\partial\Omega_S} v_n n \varphi \, ds = (R, \varphi)_{2,\Omega} - \int_{\partial\Omega_S} v_n n_D \varphi \, ds$$

for all $\varphi \in H_0^1(\Omega \cup \partial\Omega_N \cup \partial\Omega_S)$, (5.4.7) (a)

$$n|_{\partial\Omega_O} = n_D|_{\partial\Omega_O} \tag{5.4.7) (b}$$

and the corresponding piecewise linear finite element discretisation is given by:

$$-b_{n,\Delta}(\psi_\Delta; n_\Delta, \varphi_\Delta) - \int_{\partial\Omega_S} v_n n_\Delta \varphi_\Delta \, ds = (R, \varphi_\Delta)_{2,\Omega,\Delta} - \int_{\partial\Omega_S} v_n (n_D)_\Delta \varphi_\Delta \, ds$$

for all $\varphi_\Delta \in \{\varphi_\Delta \in X_\Delta | \varphi_\Delta|_{\partial\Omega_O} = 0\}$, (5.4.8) (a)

$$n_\Delta|_{\partial\Omega_O} = (n_D)_\Delta|_{\partial\Omega_O}. \tag{5.4.8) (b}$$

The boundary integrals can be evaluated exactly (assuming that $\partial\Omega_S$ is a union of line or, resp., plane segments and taking v_n piecewise constant) or approximately by using numerical quadrature.
The combined current relation-continuity equation for holes is treated analogously.

Adaptive Mesh-Construction

The performance of a steady state device simulation program depends to a great extend on its ability to devise a mesh, which generates discrete solutions of sufficient (user specified) accuracy and which has as few nodes as possible. Because of the complexity of the problem it is generally impossible to set up such a mesh a-priori. Instead, an iterative procedure is usually employed. A solution is computed on a coarse initial grid, which is set up either by taking into account available physical or/and mathematical information on the structure of solutions or, when the computation is performed within a voltage-current characteristic continuation, the final mesh of the computation with the previous data set is used as initial mesh for the next step in the continuation process (see [5.18]). Then the mesh is updated by exploiting the numerical solution and a new solution is computed on the updated mesh. Usually, the 'old' solution – appropriately interpolated – is used as initial guess for Newton's or Gummel's iteration to compute a 'new' solution. The mesh – updating iteration is stopped as soon as a computed solution is found to be sufficiently accurate.
We shall now devise an appropriate strategy for adaptive mesh-construction. From the convergence analysis of Section 5.3 we conclude that the final mesh has to be sufficiently fine in order to accurately resolve the current densities, the recombination-generation rate and the mobilities in Ω and it has to produce a discrete potential, which is accurate in outer regions. Although an accurate resolution of the potential within layer regions is not necessary and sometimes even impossible, we demonstrated that the accuracy of the potential inside the layers 'increases' (at least pointwise at the nodes) when the number of nodes located inside the layers is increased, even when the mesh is still not sufficiently fine in order to guarantee the full equidistribution of the local discretisation error of Poisson's equation. Therefore an appropriate mesh-adaptation strategy can be based on resolving the current densities, the mobilities and the recombination-generation rate accurately in Ω, approximating the potential accurately away from device-junction-, Schottky-contact- and semiconductor-oxide interface layers and placing not more than a certain number of nodes, predefined by the user and determined by the availability of computer storage and time, into the layers. This 'limited local-error-equidistribution strategy' yields a reasonable accuracy at nodes outside the layers as well as the correct structure of the solutions inside the layers.
Based on this we define the error functionals:

$$E_{1,T_l} = d_l\left(\sum_{i=1}^{k}\left(\left\|\frac{\partial J_n}{\partial x_i}\right\|^{0,T_l} + \left\|\frac{\partial J_p}{\partial x_i}\right\|^{0,T_l}\right)\right.$$
$$\left. + \left\|\frac{R}{\mu}\right\|^{1,T_l} + \left|\frac{\mu_n}{\mu_n}\right|^{1,T_l} + \left|\frac{\mu_p}{\mu_p}\right|^{1,T_l}\right) \qquad (5.4.9)\ (a)$$

$$E_{2,T_l} = \lambda^2 d_l \sum_{i=1}^{k}\left\|\frac{\partial^3\psi}{\partial x_i^3}\right\|^{0,T_l} \qquad (5.4.9)\ (b)$$

and assume that an error tolerance κ, typically between 10^{-5} and 10^{-2}, is prescribed. Then, for an already computed discrete solution, we 'evaluate' the functionals E_{1,T_l} and E_{2,T_l} in every finite element T_l by approximating the derivatives by corresponding difference quotients (using values from neighbouring elements T_j), and check the following equidistribution conditions:

$$E_{1,T_l} + E_{2,T_l} \leqq \kappa \quad \text{if} \quad T_l \cap S = \{\ \} \tag{5.4.10 (a)}$$

$$E_{1,T_l} \leqq \kappa \quad \text{if} \quad T_l \cap S \neq \{\ \}, \tag{5.4.10 (b)}$$

where S denotes the union of all layer strips occuring in the device. If (5.4.10) holds, then we leave the element T_l unchanged, if it does not hold, then we refine the mesh locally about T_l. In order to obtain at least a structural agreement of the discrete solution with the exact solution inside the layers the mesh can be additionally refined there according to the equidistribution condition:

$$E_{2,T_l} \leqq \kappa \quad \text{if} \quad T_l \cap S \neq \{\ \}, \tag{5.4.10 (c)}$$

as long as sufficient computer resources are available.
We did not prove rigorously that discrete solutions computed on a mesh, which satisfies (5.4.10), are $O(\kappa)$-approximations to an exact solution, however, the 'spirit' of the limited equidistribution approach is a consequence of the convergence analysis presented in Section 5.3. If the mesh satisfies (5.4.10) (a), (b), then we expect the discrete solutions to be $O(\kappa) + O\big((\lambda|\ln \lambda|)^{\frac{1}{q}}\big)$-approximations at least at nodes outside layer regions ($q > k$). The accuracy within layers depends on how many nodes are placed there by means of (5.4.10) (c).
The resolution of the current densities within layers is necessary for the global qualitative agreement of 'discrete' solutions with 'continuous' solutions while the resolution of the potential within layers only affects the local quantitative agreement. This feature of the discretisation is particularly apparent in MOS-transistor simulations. Typically, the current densities are small and slowly varying in outer regions *and* in the two *pn*-junction layers. The tangential component of the minority current density varies fast within the interface layer. Thus the mesh has to be sufficiently refined close to the interface in order to facilitate 'reasonable' discrete solutions. The structure of the mesh within the *pn*-junction layers only influences the local accuracy.

There are various ways to do local mesh-refinement. For one-dimensional problems a subinterval $[x_{l-1}, x_l]$ is usually split into two by introducing an additional gridpoint in the middle when the equidistribution condition fails (a more sophisticated method is presented in [5.14]). In two-or three-dimensional simplicial finite element discretisations local mesh-refinement is more complicated when the refined mesh is required to be of acute type and to satisfy the minimum angle restriction, too. We refer the interested reader to [5.31] for a lot of information and further references.
The classical way of mesh-refinement in multi-dimensional finite difference discretisations is to introduce additional grid-hyperplanes, usually again by

halfing meshsizes. For semiconductor device simulations this approach often leads to an unnecessarily fine grid in outer regions, therefore it is desirable to allow certain grid-hyperplanes to 'terminate' outside layer regions.
As an example we consider the two-dimensional situation depicted in Fig. 5.4.1.a. The y_j-gridline terminates at x_i, the point (x_{i+1}, y_j) is no grid-point.
We approximate the missing x-difference quotient by linear interpolation between the $(j+1)$st and $(j-1)$st y-level:

$$\frac{\psi_{i+1,j}-\psi_{ij}}{h_{i+1}} \approx \frac{k_j}{k_{j+1}+k_j} \frac{\psi_{i+1,j+1}-\psi_{i,j+1}}{h_{i+1}} + \frac{k_{j+1}}{k_{j+1}+k_j} \frac{\psi_{i+1,j-1}-\psi_{i,j-1}}{h_{i+1}} \tag{5.4.11}$$

and insert the right hand side into $(\Delta_2\psi)_{ij}$. This generates an additional contribution to the local discretisation error of the order of magnitude:

$$\left(k_{j+1}+k_j+\frac{k_{j+1}^2+k_j^2}{h_{i+1}+h_i} + h_{i+1}\right) \cdot \left\| \left|\frac{\partial^3\psi}{\partial x^3}\right| + \left|\frac{\partial^3\psi}{\partial x \partial y^2}\right| + \left|\frac{\partial^3\psi}{\partial x^2 \partial y}\right| \right\|_{0,[x_{i-1},x_{i+1}]\times[y_{j-1},y_{j+1}]}. \tag{5.4.12}$$

The minimum angle restriction implies that the mesh-size ratios k_{j+1}/h_{i+1} and k_j/h_i are uniformly bounded and therefore order one accuracy of the local discretisation error is retained. Terminating x-gridlines are dealt with analogously and the SG-discretisations of the combined current relations-continuity equations are modified accordingly.

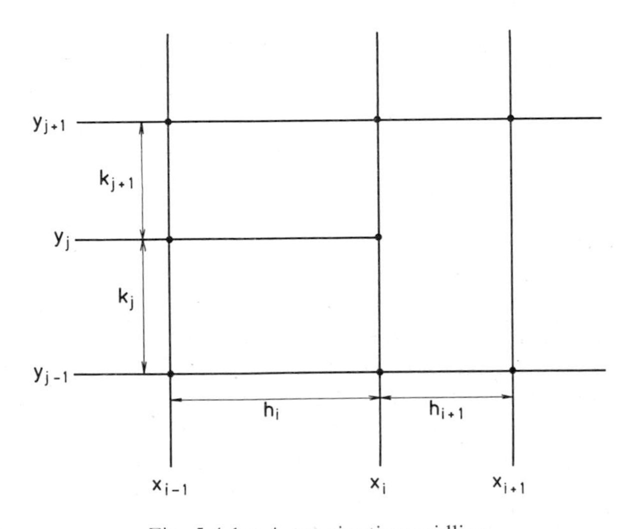

Fig. 5.4.1.a A terminating gridline

For obvious reasons gridlines should only be allowed to terminate in those regions, where the third derivatives of solutions are of moderate order of magnitude, i.e. away from layers. An algorithmic description and a practical evaluation of the terminating line approach can be found in [5.11]. It turns out that – particularly in simulations with many terminating gridlines – the saving in computer storage is paid for by comparably large condition numbers of the coefficient matrices of the discretisation due to their changed sparsity patterns. Usually, this causes a deterioration of the speed of convergence of (quasi-)linearisation methods for the coupled device problem.
These difficulties can be remedied by 'embedding' the terminating line-grid into a triangular finite element mesh (see Fig. 5.4.1.b). The finite element approximation on the 'extended' mesh reduces to a finite difference scheme in a straightforward way.
Another difficulty in many device simulations is the occurence of singularities at contact edges and at semiconductor-oxide interface edges (see Section 3.3). The error functionals E_{1,T_l} and E_{2,T_l} evaluated at the exact solution in those finite elements, which contain such edges, may be infinite, however, they have finite values when evaluated at discrete solutions in the course of the numerical computation. Since the difference quotients, which approximate the derivatives of J_n, J_p, R and ψ are usually large there, equidistribution leads to a highly refined mesh close to contact and interface edges (see [5.30]).
The terminating-line-grid, which was used for the simulation of the *pn*-diode of Fig. 4.6.1 in thermal equilibrium is depicted in Fig. 5.4.2. The grid is refined about the *pn*-junction and about contact edges. It is coarse everywhere else. The 'corresponding' finite difference grid is depicted in Fig. 5.4.3. It achieves about the same overall accuracy but consists of significantly more grid points.

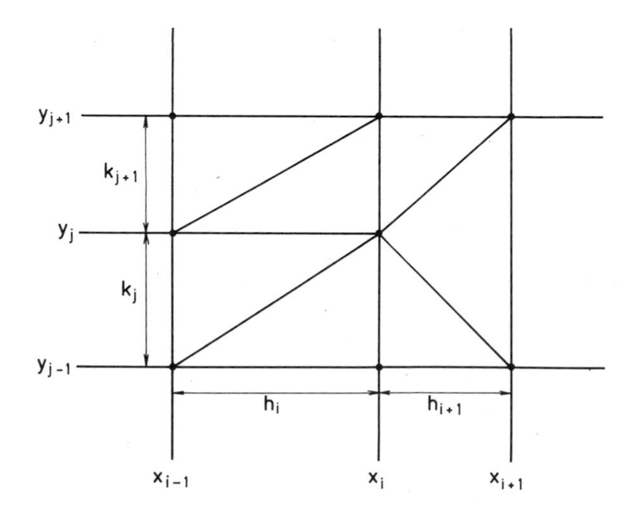

Fig. 5.4.1.b Triangular mesh

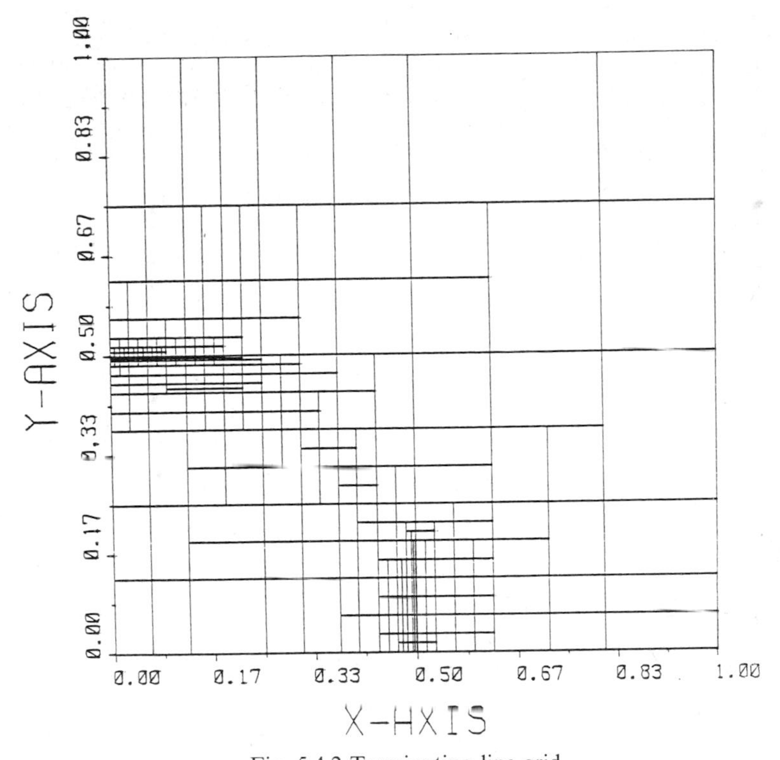

Fig. 5.4.2 Terminating-line grid

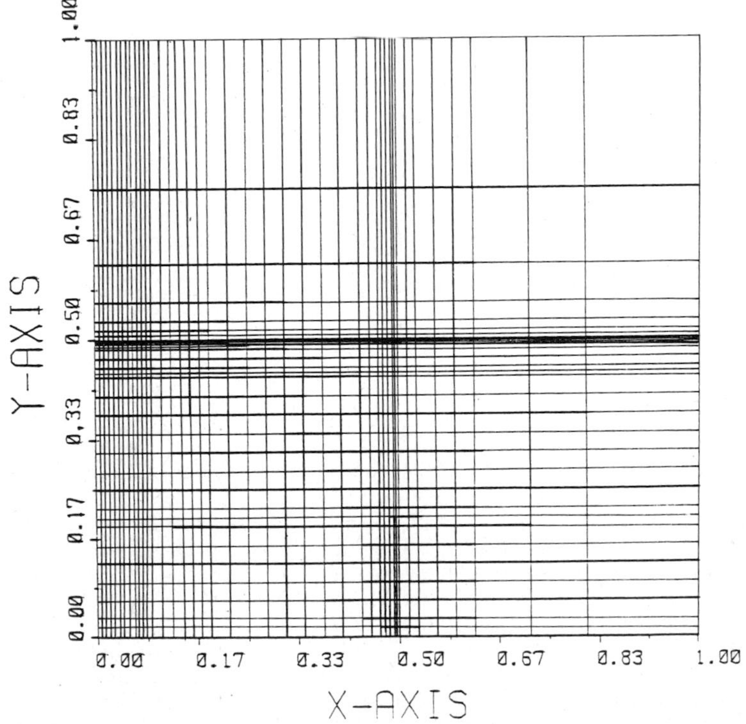

Fig. 5.4.3 Finite difference grid

References

[5.1] Axelsson, O., Barker, A. V.: Finite Element Solution of Boundary Value Problems, Theory and Computation. Orlando, Florida: Academic Press 1984.

[5.2] Bank, R. E., Jerome, J. W., Rose, D. J.: Analytical and Numerical Aspects of Semiconductor Modelling. Report 82-11274-2, Bell Laboratories, 1982.

[5.3] Bramble, J. H., Hubbard, B. E.: On the Formulation of Finite Difference Analogues of the Dirichlet Problem for Poisson's Equation. Num. Math. *4,* 313–327 (1962).

[5.4] Buturla, E. M., Cottrell, P. E.: Two-Dimensional Finite Element Analysis of Semiconductor Steady State Transport Equations. Proc. International Conference "Computer Methods in Nonlinear Mechanics", Austin, Texas, pp. 512–530 (1974).

[5.5] Choo, S. C., Seidmann, T. I.: Iterative Scheme for Computer Simulations of Semiconductor Devices. Solid State Electronics *15,* 1229–1235 (1972).

[5.6] Ciarlet, P.: The Finite Element Method for Elliptic Problems. Amsterdam–New York–Oxford: North-Holland 1978.

[5.7] Collatz, L.: Numerical Treatment of Differential Equations, 3rd ed. Berlin–Heidelberg–New York: Springer 1960.

[5.8] Doolan, E. P., Miller, J. J. H., Schilders, W. H. A.: Uniform Numerical Methods for Problems with Initial and Boundary Layers. Dublin: Boole Press 1980.

[5.9] Fichtner, W., Rose, D. J.: On the Numerical Solution of Nonlinear Elliptic PDEs Arising from Semiconductor Device Modelling. Report 80-2111-12, Bell Laboratories, 1980.

[5.10] Forsythe, G. E., Wasow, W. R.: Finite Difference Methods for Partial Differential Equations. New York: Wiley 1960.

[5.11] Franz, A. F., Franz, G. A., Selberherr, S., Ringhofer, C. A., Markowich, P. A.: Finite Boxes – A Generalisation of the Finite Difference Method Suitable for Semiconductor Device Simulation. IEEE Trans. Electron Devices. *ED–30,* No. 9, 1070–1082 (1983).

[5.12] Gilbarg, D., Trudinger, N. S.: Elliptic Partial Differential Equations of Second Order, 2nd ed. Berlin–Heidelberg–New York: Springer 1983.

[5.13] Gummel, H. K.: A Self-Consistent Iterative Scheme for One-Dimensional Steady State Transistor Calculations. IEEE Trans. Electron Devices. *ED–11,* 455–465 (1964).

[5.14] Jüngling, W., Pichler, P., Selberherr, S., Guerrero, E., Pötzl, H.: Simulation of Critical IC Fabrication Processes Using Advanced Physical and Numerical Methods. IEEE Trans. Electron Devices *ED–32,* No. 2, 156–167 (1985).

[5.15] Keller, H. B.: Approximation Methods for Nonlinear Problems with Application to Two-Point Boundary Value Problems. Math. Comp. *29,* 464–474 (1974).

[5.16] Markowich, P. A., Ringhofer, C. A.: Collocation Methods for Boundary Value Problems on 'Long' Intervals. Math. Comp. *40,* 123–150 (1983).

[5.17] Markowich, P. A., Ringhofer, C. A., Selberherr, S., Lentini, M.: A Singular Perturbation Approach for the Analysis of the Fundamental Semiconductor Equations. IEEE Trans. Electron Devices. *ED–30,* No. 9, 1165–1180 (1983).

[5.18] Markowich, P. A., Ringhofer, C. A., Steindl, A.: Arclength Continuation Methods for the Computation of Semiconductor Device Characteristics. IMA J. Num. Anal. *33,* 175–187 (1984).

[5.19] Meis, T., Marcowitz, U.: Numerische Behandlung Partieller Differentialgleichungen. Berlin–Heidelberg–New York: Springer 1978.

[5.20] Mock, M. S.: On the Convergence of Gummel's Numerical Algorithm. Solid State Electronics *15,* 781–793 (1972).

[5.21] Mock, M. S.: Analysis of Mathematical Models of Semiconductor Devices. Dublin: Boole Press 1983.

[5.22] Mock, M. S.: On the Computation of Semiconductor Device Current Characteristics by Finite Difference Methods. J. Engineering Math. *7,* No. 3, 193–205 (1973).

[5.23] Mock, M. S.: Analysis of a Discretisation Algorithm for Stationary Continuity Equations in Semiconductor Device Models I. COMPEL *2,* No. 3, 117–139 (1983).

[5.24] Mock, M. S.: Analysis of a Discretisation Algorithm for Stationary Continuity Equations in Semiconductor Device Models II. COMPEL *3,* No. 3, 137–149 (1984).

[5.25] Oden, J. T.: Finite Elements of Nonlinear Continua. New York: McGraw-Hill 1972.

[5.26] Ortega, J. M., Rheinboldt, W. C.: Iterative Solution of Nonlinear Equations in Several Variables. New York–London: Academic Press 1970.

[5.27] den Heijer, C., Rheinboldt, W. C.: On Steplength Algorithms for a Class of Continuation Methods. SIAM J. Num. Anal. *18,* Nr. 5, 925–948 (1981).
[5.28] Scharfetter, D. L., Gummel, H. K.: Large Signal Analysis of a Silicon Read Diode Oscillator. IEEE Trans. Electron Devices. *ED–16,* 64–77 (1969).
[5.29] Selberherr, S.: Analysis and Simulation of Semiconductor Devices. Wien–New York: Springer 1984.
[5.30] Strang, G., Fix, G. J.: An Analysis of the Finite Element Method. Englewood Cliffs, N. J.: Prentice-Hall 1973.
[5.31] Thompson, J. F. (ed.): Numerical Grid Generation. Amsterdam–New York–Oxford: North-Holland 1982.
[5.32] Varga, R. S.: Matrix Iterative Analysis. Englewood Cliffs, N. J.: Prentice-Hall 1962.
[5.33] Watanabe, D. S., Sheikh, Q. M., Slamed, S.: Convergence of Quasi-Newton Methods for Semiconductor Equations. Report, Department of Computer Science, University of Illinois–Urbana, U.S.A., 1984.
[5.34] Zienkiewicz, O. C.: The Finite Element Method. London: McGraw-Hill 1977.
[5.35] Zlamal, M.: Finite Element Solution of the Fundamental Equations of Semiconductor Devices I. Report, Department of Math., Technical University Brünn, CSSR, 1984.
[5.36] Zlamal, M.: A Finite Element Solution of the Nonlinear Heat Equation. RAIRO Anal. Num. *14,* 203–216 (1980).

Numerical Simulation – A Case Study 6

We shall here present numerical results from a two-dimensional simulation of a silicon MOS-transistor in order to highlight the structure of solutions of the static device problem and to demonstrate the power of the presented numerical methods and the state of the art in device modeling.
For an analysis of the performance of MOS-transistors we refer the reader to [6.3].
The computations were performed with the MOS-transistor simulation code MINIMOS, developed by S. Selberherr [6.2] and extended by A. Schütz [6.1]. The underlying model includes the SRH-, Auger- and avalanche recombination-generation rates and a velocity saturation mobility model. The finite difference method presented and analysed in Chapter 5 is used to discretise the static device problem and a Gummel-type iteration method is employed for the approximate solution of the nonlinear system of equations (see Section 3.6).
The geometry of the *n*-channel MOS-transistor is depicted in Fig. 2.3.2 and (a logarithmic plot of) the doping profile in Fig. 2.2.1. The depth of the *pn*-junction is about 0.3 μm, the lateral subdiffusion of the *n*-regions 0.7 μm, the width of the channel 1 μm and the oxide-width is about 2×10^{-2} μm. The doping profile is steeply graded at both junctions, the maximal doping concentration in the *n*-region is $\tilde{C}_n \approx 2 \times 10^{18}\ \mathrm{cm}^{-3}$ and the maximal doping concentration in the *p*-region is $\tilde{C}_p \approx 10^{15}\ \mathrm{cm}^{-3}$.
A suitable choice for the characteristic device length is $l = 1$ μm.
Since both *pn*-junctions are strongly one-sided (in the sense of Section 4.7) it is reasonable to introduce local normed Debye lengths for the *n*- and *p*-regions:

$$\lambda_n = \left(\frac{U_T \varepsilon_s}{l^2 q \tilde{C}_n}\right)^{\frac{1}{2}}, \quad \lambda_p = \left(\frac{U_T \varepsilon_s}{l^2 q \tilde{C}_p}\right)^{\frac{1}{2}}. \tag{6.1}$$

For roomtemperature $T = 300$ K we compute $\lambda_p \approx 0.04$ and $\lambda_n \approx 1.28 \times 10^{-3}$. Thus, we expect the solutions to vary significantly slower in those portions of the *pn*-layers, which are located in the *p*-region, than in the layer portions located in the *n*-regions. The variation in the *p*-region portions of the layers

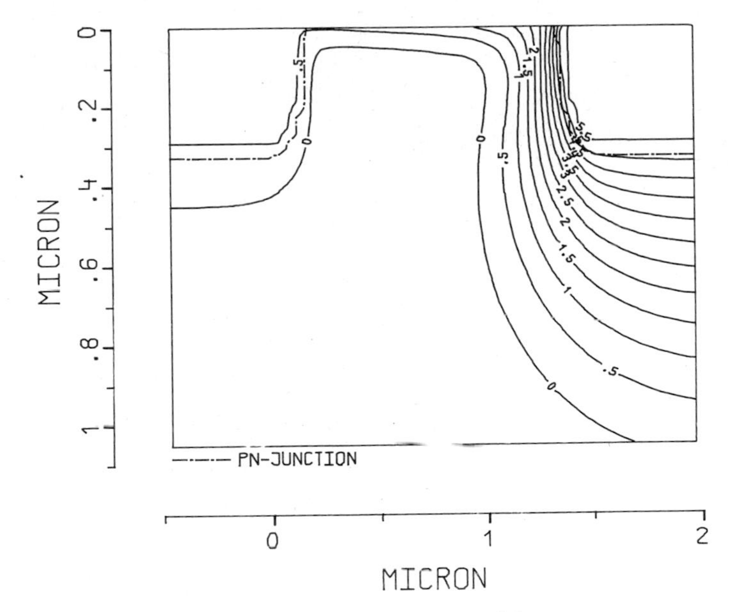

Fig. 6.1 Electrostatic potential

and in the semiconductor-oxide interface is determined by $d(x, y)/\lambda_p$ and the variation in the n-region portions of the layers by $d(x, y)/\lambda_n$, where $d(x, y)$ stands for the normal distance to the corresponding critical surface.
For the computations the operating conditions 2 V gate-source bias, 5 V drain-source bias and 0 V source-bulk bias were chosen.
A contour plot of the electrostatic potential is shown in Fig. 6.1. The pn-junction layers and the interface layer are clearly visible. As discussed above, the pn-layers stretch much further into the p-region than into the n-regions. The drain-bulk pn-junction is reverse biased, therefore the width of the layer is significantly larger than the width of the source-bulk pn-layer (note that the layer-width depends linearly on the square root of the potential drop at the junction, which is 0.5 V for the source-bulk junction and 5.5 V for the drain-bulk junction).
Fig. 6.2 and 6.3 show the electron and hole concentrations resp. The interface layer is depleted of holes and filled up with electrons. This phenomenon is – in device physics – called inversion of the channel.
The plots 6.1, 6.2, 6.3 very well demonstrate the layer-structure of ψ, n, p, i.e. the slow variation in outer regions and the fast variation within the layers.
Fig. 6.4 shows a plot of the electric field. It is large within the pn- and the interface-layers and it is small in outer regions.
The modulus of the electron current density is depicted in Fig. 6.5. The interface layer of $|J_n|$, which stems from the lateral component, is clearly visible. In fact, the layer term of the electron current density is responsible for current flow in the n-channel MOS-transistor. The two little peaks of the modulus of the electron current density come from the singularities of (the

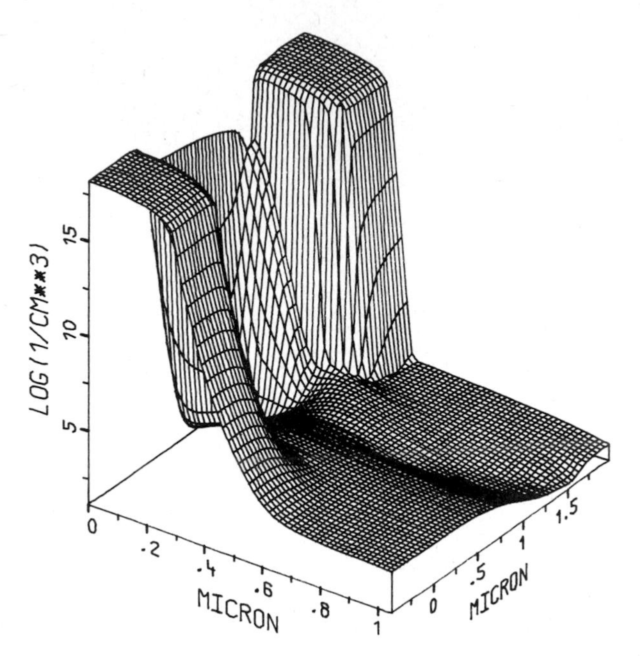

Fig. 6.2 Electron concentration [log.]

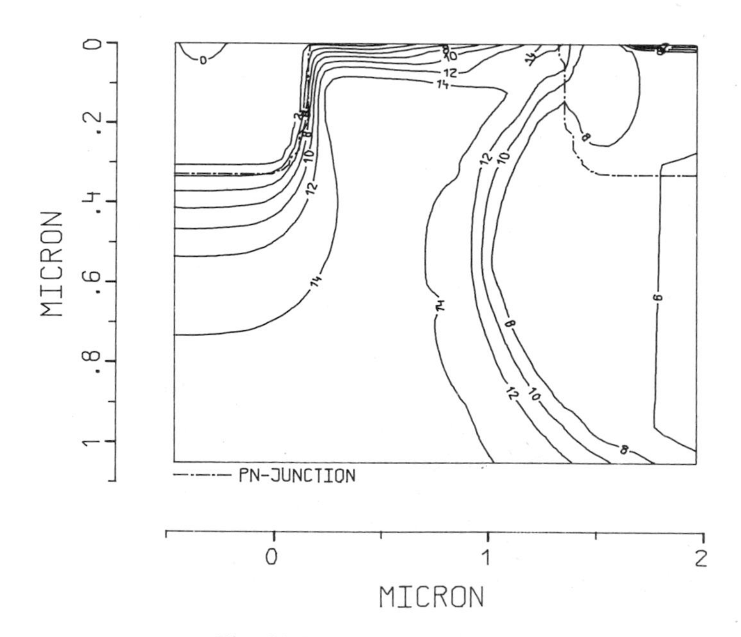

Fig. 6.3 Hole concentration [log.]

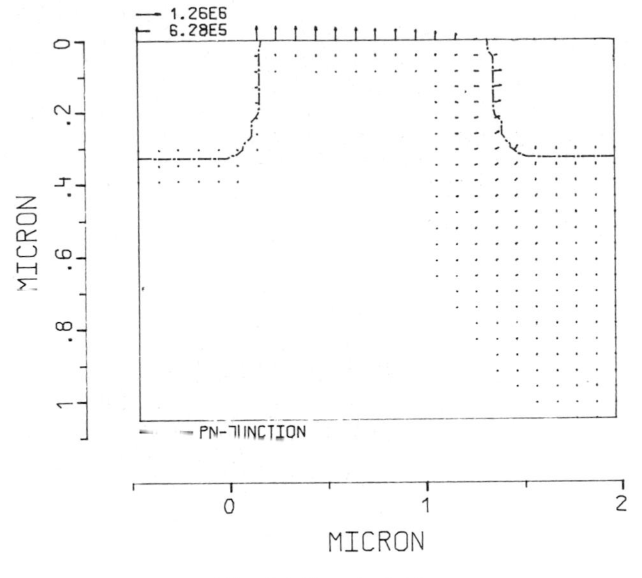

Fig. 6.4 Electric field

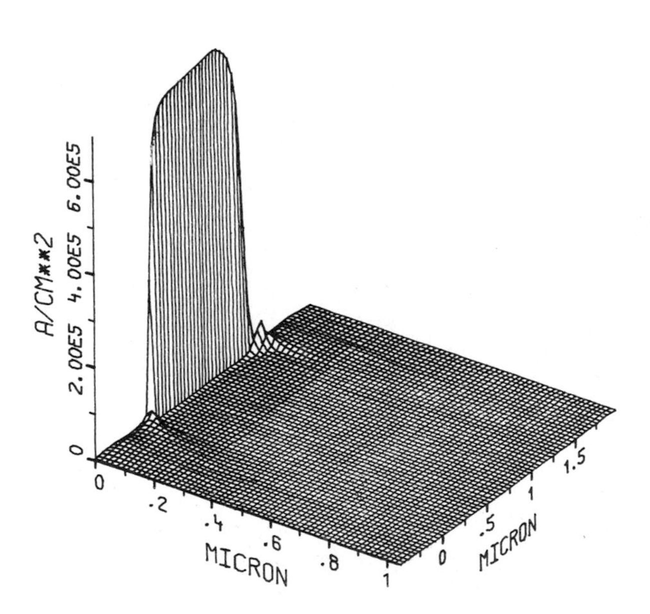

Fig. 6.5 Electron current density (modulus)

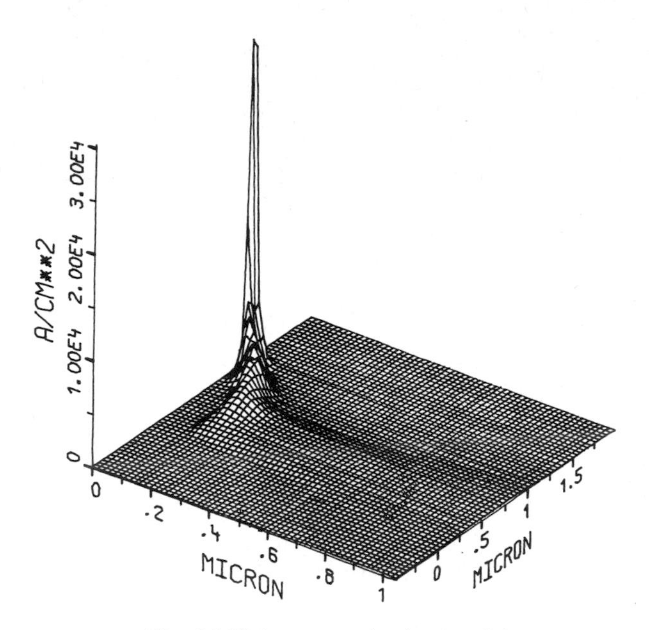

Fig. 6.6 Hole current density (modulus)

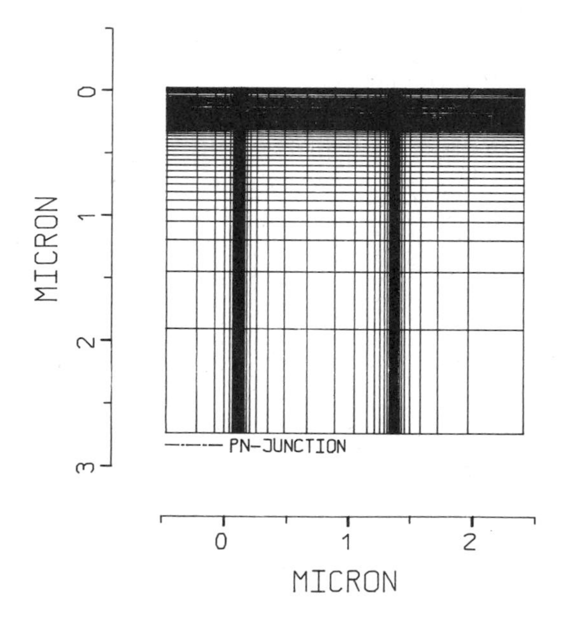

Fig. 6.7 Finite difference grid

first derivatives of) ψ, n and p at the edges of the source and drain contacts. The large peak of the modulus of the hole current density, located in the interior of the semiconductor domain, stems from a locally large avalanche generation rate (see Fig. 6.6).
The finite-difference grid, on which the discretisation was performed, is shown in Fig. 6.7. It was set up adaptively by the code by employing 'limited' local error equidistribution (see Section 5.4). The high degree of refinement in both directions about the junctions and in orthogonal direction to the interface is clearly visible.
The graphics package MOSPLOT [6.4] was used for plotting the simulation results.

References

[6.1] Schütz, A.: Simulation des Lawinendurchbruchs in MOS-Transistoren. Dissertation, Technische Universität Wien, Austria, 1982.
[6.2] Selberherr, S.: Two Dimensional Modeling of MOS-Transistors. Dissertation, TU Wien, translated by Semiconductor Physics Inc., Escondido, Cal., 1982.
[6.3] Sze, S. M.: Physics of Semiconductor Devices, 2nd ed. Cambridge–New York: J. Wiley 1981.
[6.4] Überhuber, C. et al.: MOSPLOT: A Software Package for the Graphical Presentation of MINIMOS Results. Report, Institut für Angewandte und Numerische Mathematik, Technische Universität Wien, Austria, 1985.

Appendix

Notation of Physical Quantities

B	magnetic induction vector
C	doping profile (net ionized impurity concentration)
D	electric displacement vector
D_n	electron diffusion coefficient
D_p	hole diffusion coefficient
E	electric field vector
H	magnetic field vector
J	conduction current density vector
J_n	electron current density vector
J_p	hole current density vector
N_A^-	concentration of singly ionized acceptors
N_D^+	concentration of singly ionized donors
R	carrier recombination-generation rate
R_{SRH}	Shockley–Read–Hall recombination-generation rate
R_{AU}	Auger recombination-generation rate
R_I	impact ionisation generation rate
T	device temperature
U_T	thermal voltage
α_n	electron ionisation rate
α_p	hole ionisation rate
c	speed of light
δ^2	scaled intrinsic carrier concentration
ε	permittivity
k_B	Boltzmann constant
λ_D	Debye length
λ	normed Debye length

μ_n	electron mobility
μ_p	hole mobility
m_n^*	effective electron mass
m_p^*	effective hole mass
n	electron concentration
n_i	intrinsic carrier concentration
p	hole concentration
q	elementary charge
ϱ	space charge density
t	time
τ_n^l	electron lifetime
τ_p^l	hole lifetime
τ_n^r	electron relaxation time
τ_p^r	hole relaxation time
ψ	electrostatic potential
ψ_{bi}	built-in potential
ψ_e	equilibrium potential
v_n^d	average electron drift velocity
v_p^d	average hole drift velocity
v_n^s	electron saturation velocity
v_p^s	hole saturation velocity
x	space vector

Mathematical Notation

A. Device Geometry

$k = 1, 2$ or 3	space dimension
$\Omega \subseteq \mathbb{R}^k$	semiconductor domain
$\Phi \subseteq \mathbb{R}^k$	oxide domain
$\partial\Omega$	boundary of Ω
$\partial\Omega_N \subseteq \partial\Omega$	Neumann boundary part
$\partial\Omega_D \subseteq \partial\Omega$	union of contact segments
$\partial\Omega_O$	union of Ohmic contacts
$\partial\Omega_S$	union of Schottky contacts
$O \subseteq \partial\Omega_D$	Ohmic contact
$S \subseteq \partial\Omega_D$	Schottky contact
$\partial\Omega_I \subseteq \partial\Omega$	oxide-semiconductor interface
$\Lambda = \Omega \cup \Phi \cup \partial\Omega_I$	device domain
$\bar{\Omega}$	closure of Ω
$\mathring{\Omega}$	interior of Ω
$\Gamma \subseteq \Omega$	device junction ($(k-1)$-dimensional surface), splits Ω into Ω_+, Ω_-, i.e. $\Omega = \Omega_+ \cup \Gamma \cup \Omega_-$

$\mu_k(\Omega)$	k-dimensional Lebesguemeasure of Ω, i.e. $\mu_1(\Omega)$ in the length of the interval Ω, $\mu_2(\Omega)$ the area of Ω and $\mu_3(\Omega)$ the volume of Ω
a.e. in Ω	almost everywhere in Ω with respect to $\mu_k(\Omega)$
ν	exterior unit normal vector of $\partial\Omega$
ξ	exterior unit normal vector of $\partial\Phi$

B. Scalars, Vectors and Matrices

$r \in \mathbb{R}$	real number r
$i = \sqrt{-1}$	imaginery unit
$z = z_1 + iz_2 \in \mathbb{C}$, $z_1, z_2 \in \mathbb{R}$	complex number z
$\bar{z} := z_1 - iz_2$	conjugate of z
$\lvert z\rvert := (z_1^2 + z_2^2)^{\frac{1}{2}}$	modulus of z
$a = \begin{pmatrix} a_1 \\ \vdots \\ a_m \end{pmatrix} \in \mathbb{C}^m$	(complex) m-dimensional column vector a
$a' := (a_1, \ldots, a_m)$	transposed vector of a (row vector)
$\bar{a} := \begin{pmatrix} \bar{a}_1 \\ \vdots \\ \bar{a}_m \end{pmatrix}$	conjugate vector of a
$a \cdot b = \sum_{i=1}^{m} a_i \bar{b}_i$	scalar product of the vectors a and b
$\lvert a\rvert$	a norm of the vector a
$a \times b$, $a, b \in \mathbb{R}^3$	vector product of a and b
$A = (a_{ij})_{\substack{i=1,\ldots,m \\ j=1,\ldots,l}}$	$m \times l$-matrix A with entry a_{ij} in the i-th row and j-th column
A'	transposed matrix of A
A^{-1}	inverse matrix of A
AB	product of the matrices A and B
$\lvert A\rvert$	a norm of the matrix A
I_m	$m \times m$-unit matrix
$\det A$	determinant of A

C. Functions

$f\colon \Omega \to \mathbb{R}(\mathbb{C})$	function f defined in Ω with values in $\mathbb{R}(\mathbb{C})$
$\frac{\partial f}{\partial x_i}(x) = f_{x_i}(x)$	(weak) partial derivative of f with respect to x_i at x
$f'(x)$, $x \in \Omega \subseteq \mathbb{R}$	(weak) derivative of f at $x \in \Omega \subseteq \mathbb{R}$

$\operatorname{grad} f(x) := \begin{pmatrix} \frac{\partial f}{\partial x_1}(x) \\ \vdots \\ \frac{\partial f}{\partial x_k}(x) \end{pmatrix}$	gradient of f at x
$\frac{\partial f}{\partial a}(x) := \operatorname{grad} f(x) \cdot a$	derivative of f at x in direction of the vector a
$\alpha = (\alpha_1, \ldots, \alpha_k),\ \alpha_i \in \mathbb{N} \cup \{0\}$	multiindex
$\lvert\alpha\rvert := \alpha_1 + \ldots + \alpha_k$	order of the multiindex α
$\partial^\alpha f(x) := \frac{\partial^{\lvert\alpha\rvert} f}{\partial x_1^{\alpha_1} \ldots \partial x_k^{\alpha_k}}(x)$	partial derivative of f of order $\lvert\alpha\rvert$ at x
$\Delta f(x) := \frac{\partial^2 f}{\partial x_1^2}(x) + \ldots + \frac{\partial^2 f}{\partial x_k^2}(x)$	Laplacian of f at x
$F(x) = (f_1(x), \ldots, f_k(x))',\ f_i\colon \Omega \to \mathbb{R}$	k-dimensional vectorfield on Ω
$\operatorname{div} F(x) = \frac{\partial f_1}{\partial x_1}(x) + \ldots + \frac{\partial f_k}{\partial x_k}(x)$	divergence of F at x
$\operatorname{rot} F(x) = \operatorname{grad} \times F(x) = \begin{pmatrix} \frac{\partial}{\partial x_1} \\ \frac{\partial}{\partial x_2} \\ \frac{\partial}{\partial x_3} \end{pmatrix} \times F(x)$	rotor of the three-dimensional vectorfield F
$\frac{\partial F}{\partial x}(x) = \frac{\partial(f_1, \ldots, f_k)}{\partial(x_1, \ldots, x_k)}(x) := \left(\frac{\partial f_i}{\partial x_j}(x)\right)_{\substack{i=1 \cdot k \\ j=1 \cdot k}}$	Jacobian of F at x
$f\lvert_{\Omega_1},\quad \Omega_1 \subseteq \Omega$	restriction of f to $\Omega_1 \subseteq \Omega$
$f\lvert_{\partial\Omega}$	trace of f on $\partial\Omega$
$[f(x)]_\Gamma := \lim\limits_{\substack{s \to x \in \Gamma \\ s \in \Omega_+}} f(s) - \lim\limits_{\substack{s \to x \in \Gamma \\ s \in \Omega_-}} f(s)$	'jump' of f at $x \in \Gamma$
$\int\limits_\Omega f(x)\, dx$	Lebesgue integral of f over Ω

D. Landau Symbols

a) $f(x) = O(g(x))$ for $x \to x_0$	means that $\left\lvert \frac{f(x)}{g(x)} \right\rvert \leqq$ const. for all x sufficiently close to x_0
b) $f(x) = o(g(x))$ for $x \to x_0$	means that $\lim\limits_{x \to x_0} \frac{f(x)}{g(x)} = 0.$

E. Linear Spaces of Functions and Norms

$m \in \mathbb{N} \cup \{0\}$	nonnegative integer
$C^m(\Omega)$	space of (real or complex valued) functions, which together with all their partial derivatives of order at most m are continuous in Ω
$C^\infty(\Omega)$	space of functions, whose partial derivatives of arbitrary order are continuous in Ω
$C_B^m(\Omega)$	space of those functions of $C^m(\Omega)$, which together with all their partial derivatives of order at most m are bounded in Ω
$C^m(\bar{\Omega})$	space of those functions in $C_B^m(\Omega)$, which together with all their partial derivatives of order at most m are uniformly continuous in Ω
$C(\Omega) := C^0(\Omega)$, $C_B(\Omega) := C_B^0(\Omega)$, $C(\bar{\Omega}) := C^0(\bar{\Omega})$	
$\lVert f \rVert^{m,\Omega} := \max\limits_{\lvert\alpha\rvert \leq m} \sup\limits_{x \in \Omega} \lvert \partial^\alpha f(x) \rvert$	norm on $C_B^m(\Omega)$ and on $C^m(\bar{\Omega})$
$\lvert f \rvert^{m,\Omega} := \max\limits_{\lvert\alpha\rvert = m} \sup\limits_{x \in \Omega} \lvert \partial^\alpha f(x) \rvert$	seminorm on $C_B^m(\Omega)$ and on $C^m(\bar{\Omega})$
$C^{m,\beta}(\Omega)$	space of functions, which together with all their partial derivatives of order at most m are (locally) Hölder continuous in Ω with Hölder exponent β, $0 < \beta \leqq 1$

Hölder continuity is called Lipschitz continuity for $\beta = 1$!

$C^{m,\beta}(\bar{\Omega})$	space of functions, which together with all their partial derivatives of order at most m are (globally) Hölder continuous in $\bar{\Omega}$ with Hölder-exponent β, $0 < \beta \leqq 1$
$\lVert f \rVert^{m,\beta,\Omega} := \lVert f \rVert^{m,\Omega} + \max\limits_{\lvert\alpha\rvert \leq m} \sup\limits_{x,y \in \Omega} \dfrac{\lvert \partial^\alpha f(x) - \partial^\alpha f(y) \rvert}{\lvert x - y \rvert^\beta}$	norm on $C^{m,\beta}(\bar{\Omega})$
$\lVert f \rVert_{q,\Omega} := \begin{cases} \left(\int_\Omega \lvert f(x) \rvert^q \, dx\right)^{\frac{1}{q}}, & 1 \leqq q < \infty \\ \sup\limits_{x \in \Omega} \lvert f(x) \rvert, & q = \infty \end{cases}$	q-norm of f

$L^q(\Omega) := \{f\colon \Omega \to \mathbb{R}(\mathbb{C}) \mid \|f\|_{q,\Omega} < \infty\}$	space of (real or complex valued) functions, which are q-integrable in Ω for $1 \leqq q < \infty$ and, resp., (essentially) bounded in Ω for $q = \infty$
$(f, g)_{2,\Omega} := \int_\Omega f(x)\bar{g}(x)\,dx$	$L^2(\Omega)$-inner product
$W^{m,q}(\Omega) := \{f\colon \Omega \to \mathbb{R}(\mathbb{C}) \mid \partial^\alpha f \in L^q(\Omega)$ for all α with $0 \leqq \lvert\alpha\rvert \leqq m\}$	(Sobolev) space of functions, which together with all their weak derivatives of order at most m are q-integrable on Ω
$\|f\|_{m,q,\Omega} := \left(\sum_{0 \leqq \lvert\alpha\rvert \leqq m} \|\partial^\alpha f\|_{q,\Omega}^q\right)^{\frac{1}{q}}$	norm on $W^{m,q}(\Omega)$
$\lvert f\rvert_{m,q,\Omega} := \left(\sum_{\lvert\alpha\rvert = m} \|\partial^\alpha f\|_{q,\Omega}^q\right)^{\frac{1}{q}}$	seminorm on $W^{m,q}(\Omega)$
$H^m(\Omega) := W^{m,2}(\Omega)$	
$H_0^1(\Omega) := \{f \in H^1(\Omega) \mid f\rvert_{\partial\Omega} = 0$ in the sense of $H^1(\Omega)\}$	linear subspace of $H^1(\Omega)$ consisting of functions with a tracc, which vanishes (in the weak sense) on $\partial\Omega$
$H_0^1(\Omega \cup \partial\Omega_N) := \{f \in H^1(\Omega) \mid f\rvert_{\partial\Omega - \partial\Omega_N} = 0$ in the sense of $H^1(\Omega)\}$	linear subspace of $H^1(\Omega)$ consisting of functions with a trace, which vanishes (in the weak sense) on $\partial\Omega - \partial\Omega_N$

$\partial\Omega$ is $C^{m,\beta}$, or equivalently, Ω is a $C^{m,\beta}$-domain means that Ω has the uniform $C^{m,\beta}$-regularity property (see [3.1]).

F. Functionalanalytic Notations

$(X, \|\cdot\|_X)$, $(Y, \|\cdot\|_Y)$	linear normed spaces
$T\colon X \to Y$	operator T defined on X with values in Y
$C(X, Y)$	set of continuous operators $T\colon X \to Y$
$L(X, Y)$	set of linear continuous operators $T\colon X \to Y$
$\|T\|_{X\to Y} := \sup_{x \neq 0} \dfrac{\|Tx\|_Y}{\|x\|_X}$	norm on $L(X, Y)$
$d_x T(x; h)$	Gateaux-derivative of T at x in direction h
$D_x T(x)$	Frechet-derivative of T at x
$C^m(X, Y)$	set of m-times continuously Frechet-differentiable operators $T\colon X \to Y$
$X^* := \begin{cases} L(X, \mathbb{R}) & \text{if } X \text{ is a 'real' space} \\ L(X; \mathbb{C}) & \text{if } X \text{ a 'complex' space} \end{cases}$	dual space of X

Subject Index

Druck: Ernst Becvar, A-1150 Wien

Analysis and Simulation of Semiconductor Devices

By Dipl.-Ing. Dr. **Siegfried Selberherr,**
Institut für Allgemeine Elektrotechnik und Elektronik,
Technische Universität Wien, Austria

1984. 126 figures. XIV, 294 pages.
ISBN 3-211-81800-6

Contents: Introduction. – Some Fundamental Properties. – Process Modeling. – The Physical Parameters. – Analytical Investigations About the Basic Semiconductor Equations. – The Discretization of the Basic Semiconductor Equations. – The Solution of Systems of Nonlinear Algebraic Equations. – The Solution of Sparse Systems of Linear Equations. – A Glimpse on Results. – References. – Author Index. – Subject Index.

Numerical analysis and simulation has become a basic methodology in device research and development. This book satisfies the demand for a thorough review and judgement of the various physical and mathematical models which are in use all over the world today. A compact and critical reference with many citations is provided, which is particularly relevant to authors of device simulation programs. The physical properties of carrier transport in semiconductors are explained, great emphasis being laid on the direct applicability of all considerations. An introduction to the mathematical background of semiconductor device simulation clarifies the basis of all device simulation programs. Semiconductor device engineers will gain a more fundamental understanding of the applicability of device simulation programs. A very detailed treatment of the state-of-the-art and highly specialized numerical methods for device simulation serves in an hierarchical manner both as an introduction for newcomers and a worthwhile reference for the experienced reader.

Springer-Verlag Wien New York